Improvement of the Ozonation Process Through the Use of Static Mixers

The mission of the Awwa Research Foundation (AwwaRF) is to advance the science of water to improve the quality of life. Funded primarily through annual subscription payments from over 1,000 utilities, consulting firms, and manufacturers in North America and abroad, AwwaRF sponsors research on all aspects of drinking water, including supply and resources, treatment, monitoring and analysis, distribution, management, and health effects.

From its headquarters in Denver, Colorado, the AwwaRF staff directs and supports the efforts of over 500 volunteers, who are the heart of the research program. These volunteers, serving on various boards and committees, use their expertise to select and monitor research studies to benefit the entire drinking water community.

Research findings are disseminated through a number of technology transfer activities, including research reports, conferences, videotape summaries, and periodicals.

Improvement of the Ozonation Process Through the Use of Static Mixers

Prepared by:
Chandra Mysore and **Jerome Leparc**
Vivendi Water North American Technology Center
Norcross, GA 30092
Daniel W. Smith, Steve Craik, and **Gordon Finch**
Department of Civil and Environmental Engineering
University of Alberta, Alberta T6G 2M8
Richard Lake and **Paula Agutter**
Vivendi Water Partnership
Watford, Herts WD2 2LG – United Kingdom
Christelle de Traversay
Vivendi Water – Anjou Recherche, Maisons Laffitte, France
Dominique Wolbert
University of Rennes, France
Gary L. Amy
Department of Civil Engineering
University of Colorado, Boulder, CO 80309

Jointly Sponsored by:
Awwa Research Foundation
6666 West Quincy Ave.
Denver, CO 80235-3098

and

Vivendi Water

Published by:

Awwa Research Foundation
Advancing the Science of Water®

IWA Publishing

DISCLAIMER

This study was jointly funded by the Awwa Research Foundation (AwwaRF) and by Vivendi Water under a Tailored Collaboration Research Project No. 2537. AwwaRF and Vivendi Water assume no responsibility for the content of the research study reported in this publication or for the opinions or statements of fact expressed in the report. The mention of trade names for commercial products does not represent or imply the approval or endorsement of AwwaRF or Vivendi Water. This report is presented solely for informational purposes.

Published by IWA Publishing, Alliance House, 12 Caxton Street, London SW1H 0QS, UK

Telephone: +44 (0) 20 7654 5500; Fax: +44 (0) 20 7654 5555; Email: publications@iwap.co.uk;
Web: **www.iwapublishing.com**

AwwaRF report number 90930F. Originally published by AwwaRF for its subscribers in 2003.
IWA Publishing version published 2004.

ISBN 1 84339 848 6

CONTENTS

TABLES

FIGURES

FOREWORD

The Awwa Research Foundation is a nonprofit corporation that is dedicated to the implementation of a research effort to help utilities respond to regulatory requirements and traditional high-priority concerns of the industry. The research agenda is developed through a process of consultation with subscribers and drinking water professionals. Under the umbrella of a Strategic Research Plan, the Research Advisory Council prioritizes the suggested projects based upon current and future needs, applicability, and past work; the recommendations are forwarded to the Board of Trustees for final selection. The foundation also sponsors research projects through the unsolicited proposal process; the Collaborative Research, Research Applications, and Tailored Collaborations programs; and various joint research efforts with organizations such as the U.S. Environmental Protection Agency, the U.S. Bureau of Reclamation, and the Association of California Water Agencies.

This publication is the result of one of these sponsored studies, and it is hoped that its findings will be applied in communities throughout the world. The following report serves not only as a means of communicating the results of the water industry's centralized research program but also as a tool to enlist the further support of the nonmember utilities and individuals.

Projects are managed closely from their inception to the final report by the foundation's staff and large cadre of volunteers who willingly contribute their time and expertise. The foundation serves a planning and management function and awards contracts to other institutions such as water utilities, universities, and engineering firms. The funding for this research effort comes primarily from the Subscription Program, through which water utilities subscribe to the research program and make an annual payment proportionate to the volume of water they deliver and consultants and manufacturers subscribe based on their annual billings. The program offers a cost-effective and fair method for funding research in the public interest.

A broad spectrum of water supply issues are addressed by the foundation's research agenda: resources, treatment and operations, distribution and storage, water quality and analysis, toxicology, economics, and management. The ultimate purpose of the coordinated effort is to assist water suppliers to provide the highest possible quality of water in an economical and reliable manner. The true benefits are realized when the results are implemented at the utility level. The foundation's trustees are pleased to offer this publication as a contribution toward that end.

Interest in ozonation in water treatment within United States has increased substantially in the 1990s because of suspected carcinogenic properties of trihalomethanes that are formed when organic compounds, naturally present in water, react with chlorine. Among the known disinfectants, ozone has been demonstrated to be very effective in inactivating protozoan cysts and oocysts. Current and proposed regulations impose additional treatment requirements for *Cryptosporidium parvum*. These regulatory trends tend to place more stringent performance demands on disinfection systems and have, therefore, increased the need for improvements in design. The static mixer offers one alternative method for improving the efficiency of the dissolution of ozone gas and for optimizing the use of ozone in the disinfection process. This report documents the benefits of using a static mixer for ozone dissolution in drinking water treatment. Supported by comprehensive bench-, pilot- and full-scale studies under a variety of operating conditions and source waters, both in North America and the Europe, this report provides valuable information and recommendations to optimize the disinfection performance of the ozonation treatment while limiting bromate formation.

Edmund G. Archuleta, P.E.
Chair, Board of Trustees
Awwa Research Foundation

James F. Manwaring, P.E.
Executive Director
Awwa Research Foundation

ACKNOWLEDGMENTS

The authors of this report would like to thank the following water utilities, individuals, government agencies, and private businesses for their cooperation and participation in this project:

Gwinnett County Department of Public Utilities, GA; Neil Spivey
City of Cleveland, Division of Water, OH; Maggie Rogders and Don Zagorski
Elizabethtown Water Company, NJ; Edward Mullen
Los Angeles Department of Water and Power, CA; Gary Stolarik and Don Christie
Vivendi Water Partnership, United Kingdom, Terry Burke and Richard Lake

The authors also wish to express their appreciation to the Awwa Research Foundation for funding this research: Kenan Ozekin, Project Manager, and the Project Advisory Committee for their helpful comments, suggestions and advice: Appiah Amirtharajah of the Georgia Institute of Technology, Benito Marinas of the University of Illinois-Urbana Champaign, Carol Ashe of Camp Dresser & McKee and Laura Cummings of Passaic Valley Water Commission. Special thanks go to Stewart Mehlman from Praxair Trailigaz for providing an ozone generator and ancillary equipment for the pilot unit.

EXECUTIVE SUMMARY

The recently promulgated Interim Enhanced Surface Water Treatment Rule (IESWTR) requires 2 Log removal of *Cryptosporidium parvum* in drinking water treatment plants (WTPs). The proposed Long Term 2 Enhanced Surface Water Treatment Rule (LT2ESWTR) places additional treatment requirements (up to 2.5 Log reduction) for WTPs employing conventional treatment that are in full compliance with the IESWTR. These additional treatment requirements are based on source water characterization of average oocyst occurrence. These regulatory trends tend to place more stringent performance demands on disinfection systems and have, therefore, increased the need for improvements in the design of such systems.

The static mixer offers one alternative method for improving the efficiency of the dissolution of ozone gas and for optimizing the use of ozone in the disinfection process. A static mixer consists of a series of stationary mixing elements inserted end-to-end along the direction of flow in a pipe. Each of the mixing elements is a rigid structure which divides and recombines the flow stream. The fluid is directed through the flow channels provided by the mixing elements. The fluid becomes more homogeneous along the length of the elements. The introduction of static mixing elements not only accelerates radial mixing but also produces a higher and more uniform rate of turbulent energy dissipation, thereby promoting micromixing.

OBJECTIVES

The overall objective of this research was to quantify the potential benefits of ozone application through the use of static mixers in terms of increased transfer efficiency, disinfection capacity, and enhanced chemical reactions at laboratory-scale, pilot-scale and full-scale systems. Specific goals were to identify and quantify the effect of several water quality parameters (e.g., pH, DOC, alkalinity, etc) and environmental/engineered factors (e.g., temperature, transferred ozone dose, and liquid velocity) on the disinfection capacity of the ozone-static mixer system and to assess bromate formation under optimum conditions for microbial inactivation.

APPROACH

An in-depth comprehensive approach comprising of laboratory- and pilot-scale testing, supported by data from full-scale plants was formulated to meet the goals of this research project. Laboratory-scale tests studied inactivation of regulated microbial contaminants (*Cryptosporidium parvum*) and surrogates (*Bacillus subtilis* spores and *Giardia muris*) over a range of pH and temperature conditions. Pilot-tests at six sites assessed inactivation of microbial surrogates (*Bacillus subtilis*), destruction of pesticides, and formation of bromate over a range of natural and spiked water quality conditions. Pilot-scale tests also focused on side-by-side process performance comparisons of static mixer systems and fine bubble diffusion (FBD) contactor systems, as well as provided data for modeling reactor hydrodynamics, process performance of static mixers, and bromate formation. Data from full-scale plants fitted with static mixers in the UK for ozonation were used to confirm findings of the laboratory- and pilot–scale testing as well as to verify hydraulic and kinetic models. The focus of the investigations was inactivation of microbial contaminants and surrogates, destruction of pesticides and micropollutants, and occurrence of by-products such as bromate formation.

To address the issue of scale-up, to ensure that the bench- and pilot-scale studies be designed to best represent full-scale treatment plants, and the results from three phases complement one another, the following experimental plan was developed:

1. Two testing units, one for lab-scale (flow = 7.6 to 11.4 L/min, 2 to 3 gpm) and one for pilot-scale (flow = 3.8 to 38 L/min, 1 to 10 gpm) were built simultaneously at the University of Alberta to conduct side-by-side studies and address the issue of scale-up. Laboratory and pilot studies were done with a 15.7 mm (0.6 inch) diameter static mixer. A major portion of the work was conducted with *Bacillus subtilis* spores in bench- and pilot-scale studies. A few trials with *Giardia muris* and *Cryptosporidium parvum* were conducted in the subsequent phases of the lab-scale study to test the design predictions for the apparatus.
2. To have meaningful comparisons between pilot-scale and full-scale results, side-by-side testing was conducted in the U.K. at the Chertsey WTP with the pilot unit operated next to a full-scale static mixer. Native aerobic spore-formers and *Bacillus subtilis* spores were also monitored at full-scale. The inactivation of *Bacillus subtilis* spores by ozone formed the basis for comparing the three scales of process examined in this project.
3. Full-scale testing in the U.K. studied microbial inactivation, pesticide/micropollutant destruction and bromate formation under typical operating conditions at four water treatment plants (WTPs) fitted with static mixers for ozonation. Data collected from the three testing phases was processed and used in the modeling.

The project sponsors were AwwaRF and Vivendi Water (VW). Also contributing was Vivendi Water Partnership (VWP) in the United Kingdom (UK). The laboratory-scale study was done by the University of Alberta (UOA) in Canada and VW conducted the pilot-studies in the US and UK. Hydrodynamic and mass transfer modeling was done by the University of Rennes in France and bromate formation modeling was done by the University of Colorado. Participating utilities were Los Angeles Department of Water and Power, CA; Elizabethtown Water Company, NJ; Cleveland Division of Water, OH; Gwinnett County, GA; Lake Manatee Water Treatment Plant, FL and VWP in the U.K., all of which hosted pilot demonstrations and presented a variety of different surface waters. For full scale demonstrations, VWP in the UK was the utility contributor, and additional pilot work was sponsored by Suburb of Paris, France (VW).

REPORT ORGANIZATION

The results obtained from lab-scale (Chapter 2); pilot-scale (Chapter 3); and full-scale studies (Chapter 4) over a wide variety of operating conditions and source waters are presented to improve our understanding of the disinfection process in static mixers while using ozone as a disinfectant. Data on transfer efficiency, microbial inactivation, bromate formation, and destruction of micropollutants due to ozonation via a static mixer are also presented. The results were used to develop an hydrodynamic model (Chapter 5) and a bromate formation model (Chapter 6) for static mixers. Major conclusions from this research effort are presented in the next section.

CONCLUSIONS

Static Mixers for Ozone Dissolution

Ozone dissolution performance of static mixers were studied at lab-, pilot- and full-scale. Based on the results obtained, the following conclusions can be drawn:

1. The ozone transfer efficiency is primarily influenced by static mixer hydrodynamic parameters i.e. superficial liquid velocity, v_s, and the gas-liquid flow rate ratio, G/L, within the static mixer. The ozone dose and the ozone residual at the outlet of the static mixer have little influence on the transfer efficiency compared to v_s or G/L.
2. Transfer efficiency was greatest at maximum v_s and minimum G/L. For the waters tested, a G/L ratio of 0.6 to 1.5 % and a liquid velocity of 0.7 to 1.6 m/sec resulted in the highest transfer efficiency (96% to 99%). Beyond this velocity, there was no further improvement of transfer efficiency. This velocity corresponded to a pressure drop of 10 to 20 KPa (1.5 to 2.9 psi).
3. A mathematical model was developed that related transfer efficiency to the superficial liquid velocity, v_s, and the gas-liquid flow rate ratio, *G/L*, within the static mixer. Modeling of the results showed that 94 % of the ozone transfer occurred in the static mixer itself and the gas-liquid separator provides little additional transfer. The static mixer was capable of achieving close to maximum theoretical ozone mass transfer.
4. Data collected from full-scale plants utilizing static mixers showed that transfer efficiencies of 95 to 98% were achieved during regular operations. Transfer efficiencies greater than 99% were recorded at the Clay Lane plant in the UK, which employs static mixers in the main line and sidestream.

Inactivation of Microorganisms and Bromate Formation

Inactivation of *Bacillus subtilis*, *Cryptosporidum parvum* and *Giardia muris* were studied at laboratory-scale. At pilot-scale, inactivation of *Bacillus subtilis* was studied. Inactivation of *Bacillus subtilis* spores, indigenous aerobic spores and coliforms were studied at full-scale. Bromate formation was studied at pilot- and full-scale. The following conclusions are drawn from these studies:

1. The residence time distribution of the lab-scale and pilot–scale units was characteristic of a plug flow with dispersion type hydraulic regime and no short-circuiting. This hydraulic characteristic was used to achieve efficient microbial inactivation in disinfection experiments. The lab-scale results showed that an integrated ozone disinfection process consisting of a static mixer, gas-liquid separator and a plug flow contactor can achieve microbial inactivation that is comparable to an idealized well-mixed batch reactor or plug flow reactor given the same average ozone concentration and contact time. Disinfection results obtained in the lab-scale contactor can be used to predict disinfection in the pilot-scale contactor. The disinfection process is scale-able. Kinetic results obtained from lab-scale reactors can be used to predict performance in larger continuous-flow ozone contactors.

2. No significant microbial inactivation was observed at the outlet of the static mixer. In challenge experiments with *Bacillus subtilis* spores, *Cryptosporidium parvum* oocysts and *Giardia muris* cysts at lab-scale, there was little evidence of significant microorganism inactivation due to direct gas-liquid ozone contact within the static mixer itself.
3. Although no microbial inactivation was measured within the static mixer itself, the static mixer provided the necessary starting conditions for efficient inactivation in the overall contacting system. These were (i) rapid and efficient dissolution of gaseous ozone and (ii) generation of a homogeneous dissolved ozone solution. The rapid and efficient dissolution of gaseous ozone lead to better Ct values thereby effectively exploiting the volume of the contactor. The generation of a homogeneous dissolved ozone solution was possible due to the micromixing environment offered by the static mixer. The lab study also demonstrated that a static mixer, when integrated into a well-designed, continuous-plug flow ozone contacting system, can achieve excellent and predictable *Cryptosporidium parvum* inactivation. In such a system, *Cryptosporidium parvum* inactivation can be predicted using relatively straightforward modeling techniques. Measured *Cryptosporidium parvum* inactivation was equivalent to (or slightly better) than that predicted using the published kinetic model of inactivation by ozone assuming plug flow hydrodynamics.
4. Results from pilot-scale studies conducted at the various sites showed that higher ozone residuals (compared to a FBD unit) were obtained in a short interval of time at the outlet of the static mixer due to intense mixing and ozone transfer through the static mixer elements. Enhanced microbial inactivation was observed due to a consequence of the higher Ct values reached along the static mixer unit.
5. A comparison of the results from pilot-scale studies conducted at the various sites in terms of ozone dissolution performance, microbial inactivation and bromate formation showed that water quality parameters (temperature and pH among other factors) influence disinfection efficiency and bromate formation. The effect of temperature on inactivation of *Bacillus subtilis* with the pilot-scale unit confirmed the results obtained with the laboratory-scale unit and the well-stirred batch reactor. To achieve comparable inactivation, higher Ct values are required at cold temperatures. High disinfection efficiency and low bromate formation were observed at low pH (<7.0). The results underscored the importance of ozone dose, pH and alkalinity on bromate formation.
6. The Ct values for *Bacillus subtilis* and *Cryptosporidium parvum* obtained during laboratory-scale, pilot-scale studies, modeling and from literature were compared at different temperatures. The comparison indicated that at higher temperatures (≥ 16°C), the Ct requirements for similar levels of inactivation of both microorganisms are comparable. At temperatures 12°C to 3°C, *Bacillus subtilis* is inactivated more readily than *Cryptosporidium parvum.* The comparison of Ct values at these temperatures showed that 2-log inactivation of *Cryptosporidium parvum* can be achieved at 2-2.3 times the equivalent *Bacillus subtilis* Ct value. The modeling effort showed that Ct requirements for 1 log-unit inactivation of *Cryptosporidium parvum* or *Bacillus subtilis* are similar at all temperatures. On comparing Ct values for 2 log inactivation of *Cryptosporidium* (Finch, 1999) at high and low temperatures with the Cts obtained from this study either thru lab work or modeling, it was found that the Ct required was

reduced by 42-52%. Similar trends were seen while comparing Ct values derived in natural waters (Oppenheimer, 1999) and this work. This could have serious implications on the development of Ct tables for inactivation of *Cryptosporidium parvum*.

The SWTR Guidance Manual provides technical guidance for the design of surface water treatment facilities (1990). The guidance manual focuses on the inactivation of *Giardia* and viruses, and Ct values for ozone inactivation of *Giardia* are provided in Appendix O. No Ct values are specified for the inactivation of *Cryptosporidium*; however, a commonly used method for determining *Cryptosporidium* inactivation has been to apply a linear multiplier to the tabulated *Giardia* Ct values to determine the *Cryptosporidium* Ct. In the absence of guidelines, ozone system designers have typically used multipliers ranging from 5 to 20 times *Giardia* Ct values to determine Cryptosporidium inactivation Ct values. In comparison, the potential for *Bacillus subtilis* to serve as a surrogate for *Cryptosporidium parvum* appears to be much better. Disinfection studies can be complex to design and construct, especially when pathogenic protozoa (i.e. *Cryptosporidium* and *Giardia*) are included. In such studies and other performance-based testing, where quantifying the inactivation of target microorganisms (e.g. *Cryptosporidium*) are required, *Bacillus subtilis* will serve as a very useful surrogate as they are easy to culture and enumerate. *Bacillus subtilis* spores are also useful for evaluation of the hydrodynamic efficiency of alternative ozone contactor designs, as they were found to be relatively resistant to ozone over a large temperature and pH range.

7. The results from full-scale studies showed that the inactivation of naturally occurring aerobic spores and *Bacillus subtilis* varied between 0.2 and 2.2 log. In general, *Bacillus subtilis* was less resistant to ozone than the other microorganisms tested. An optimal ozone dose of 0.2 to 0.4 mg O_3/mg C ensured bromate below the MCL (10 ppb) at all times even though the pH varied between 7.2 and 8.3. Bromate formation was shown to be influenced by the contact time, ozone dose, alkalinity, pH and the level of bromide in the inlet water. pH was the most influential parameter that controlled bromate formation.
8. In general, applying low ozone doses ($\leq$ 0.5 mg O_3/mg C) through a static mixer ozone dissolution system integrated with a well designed downstream contactor (plug flow) combined with a pH ($\leq$ 7.0) can provide adequate microbial inactivation while keeping bromate formation below the current MCL (10 ppb). For some waters, it may be necessary to depress the pH to 6.5 to achieve similar results. The most challenging waters to treat will be extremely cold waters with significant concentrations of bromide.

Static Mixers Versus Fine Bubble Diffusion Contactors

Pilot studies were conducted at the Los Angeles Department of Water and Power, CA (LADWP) and Elizabethtown Water Company, NJ (EWC) where the static mixer unit (SMU) was compared side-by-side with a fine bubble diffusion (FBD) pilot unit. Data from a full-scale plant in the UK (Iver) which employed FBD contactors and turbine diffusers was also garnered to compare their effectiveness against plants that employed static mixers for ozone dissolution. The following conclusions are made based on the results of these studies:

1. The results of this study showed that intense mixing and ozone transfer through the static mixer elements leads to a much faster ozone dissolution and systematically higher Ct values along the SMU (1.2-2.3 times) than those provided through the FBD unit of equivalent contact time. Inactivation data obtained with *Bacillus subtilis* at LADWP and EWC showed that the SMU provided enhanced inactivation (0.4 to 2.9 Log more than FBD unit) under similar applied ozone doses, water quality conditions, and equivalent contact times. This improvement was due to higher Ct values reached along the SMU. While utilizing the SMU at the LADWP, higher microbial inactivation was achieved at an adjusted pH of 6.5 than the natural water pH due to a higher Ct obtained at the lower pH. Also, bromate formation was minimized at the lower pH (6.5). For cold waters, such as the one encountered in EWC (5°C), higher Ct was necessary to achieve comparable levels of microbial inactivation. No significant difference in bromate formation was observed between SMU and FBD units while testing them at LADWP and EWC.
2. The results from full-scale studies showed that static mixers are more effective for inactivation of microorganisms than conventional diffusers, particularly for the inactivation of coliforms. The inactivation at the Iver plant was significantly lower as compared to plants that employ static mixers. Also, the lowest inactivation of aerobic spores and *Bacillus subtilis* spores was observed at Iver.

Hydrodynamic and Bromate Formation Modeling

Hydrodynamic modeling was performed to model the ozone transfer and consumption through the static mixer and the gas-liquid separator utilizing bench and pilot data incorporating natural water quality conditions. A satisfactory simulation was obtained for the ozone concentration (residual) at the outlet of the static mixer or of the gas/liquid separator, ozone consumption rate and transfer efficiency. The model simulated the transfer and reactivity of ozone in static mixers for various configurations, as well as the contribution of the gas/liquid separator quite well.

Bromate formation models were developed incorporating the following parameters: bromide concentration, ozone dose, pH, temperature, DOC, contact time and alkalinity. The simulations for the SMU showed that predicted bromate was in agreement with measured bromate.

Advantages of Static Mixers for Ozonation

Based on the data gathered at bench-, pilot- and full-scale plants from this project and the operational experience of full-scale plants employing static mixers in Europe, the adavantages of using static mixers for ozonation can be summarized as follows:

1. Static mixers exhibit very high transfer efficiencies (up to 99% were recorded in tests in the UK full-scale plants; Davidson, 1997), are very effective in destroying pesticides such as atrazine, inactivating microorganisms, removal of color and eliminating taste and odor problems, all within a few seconds. The improvement in

transfer efficiencies leads to lower applied ozone dosages to achieve equivalent Ct conditions, leading to lower operational costs.

2. In a static mixer, the mass transfer coefficient (k_La) ranges from 0.1 to 10 s^{-1}. Transfer coefficient of about 5 s^{-1} for 400 W/kg of dissipated power have been reported (Middleton, 1978), which is 100 to 1000 times greater than for a baffled reactor or a classical gas/liquid separator (Hérès, 1993). The results from this study suggested that for a given ozone concentration, the most efficient operation is at a low G/L and high feed gas ozone concentration. Improved transfer efficiency at low gas flows is an advantage of static mixers because it means that these devices can be effectively integrated with modern, high-concentration, medium frequency ozone generators. In comparison, conventional fine bubble diffusion systems are limited by a requirement of sufficient gas flow rate to ensure adequate gas-liquid mixing and ozone dissolution (Schulz et al. 1995).
3. The results from this study have important implications for the design of ozone contactors with static mixers in full-scale plants. The FBD systems require large ozone injection/transfer chambers for ozone dissolution. The results of this study showed that intense mixing and ozone transfer through the static mixer elements leads to a much faster ozone dissolution and systematically higher Ct values along the SMU than those provided through the FBD unit. This study also showed that Ct requirements could be lowered by 42% to 52% to achieve the same level of Cryptosporidium inactivation. The utilization of static mixers as an ozone dissolution technology can lead to reduction in the contactor size downstream of the ozone dissolution system as targeted inactivation efficiencies are reached more quickly with an optimized Ct as compared to FBD systems. As the hydraulic residence time (HRT) of the static mixer is very small (few seconds), the overall footprint of the ozonation process is also significantly reduced when compared to a FBD system which needs a large injection chamber for ozone dissolution. This overall reduction in HRT (25 to 45% reductions were seen in this study) for the ozone contactors results in significant savings in capital, operation and maintenance (O & M) costs.
4. Static mixer systems produce a homogeneous ozone solution, have no moving parts, are easy to retrofit, require minimal maintenance, have low capital and O & M costs, have low energy consumption, possess wide turndown ratios, are adaptable to automatic process control, are simple to operate, and have no small holes which could be gradually plugged. Headlosses in the static mixer system can be significantly reduced by adopting a sidestream option to a foot or less (Chapter 4 and Figure B.1 of Appendix B). The porous plates of FBD systems have small holes which get plugged (iron or/and manganese) requiring frequent maintenance, the porous plates themselves have to be replaced every 5 years and the rubber gaskets are subject to fouling/wear and tear. Table B.5 (Appendix B) presents a cost comparison for a typical drinking WTP with the static mixer versus FBD system option for ozone dissolution.

RECOMMENDATIONS FOR FUTURE WORK

The work done in this project helped identify several important areas for future research that are of value to the drinking water industry. Specific recommendations are listed below:

1. Inactivation of *Cryptosporidium parvum* oocysts by ozone were conducted with the laboratory study static mixer ozone contactor at the feed water temperature of 22°C. More inactivation data needs to be developed at cold water temperatures especially in continuous flow systems such as the one used in this study. This data needs to be compared to inactivation data obtained for *Bacillus subtilis* (also at cold temperatures) to develop the equivalent Ct value for inactivation of *Cryptosporidium parvum*. These experiments are necessary as spore inactivation characterisitics are sensitive to artifacts such as clumping and agglomeration and may vary from batch to batch. This would help in further confirming the suitability of *Bacillus subtilis* as a surrogate to *Cryptosporidium parvum*. Also future research should be done using oocysts that have been exposed to environmental conditions that are representative of a real world system, rather than using "fresh" laboratory preserved oocysts.
2. The bench and pilot studies should compare the performance of different types of static mixers. The effects of other static mixer design parameters like the number of mixing elements and the diameter of the static mixer need to be studied.
3. More tests need to be done to study the effect of ammonia in lowering bromate formation. Future experimental work should investigate different injection points for introducing ozone in the static mixer as some studies have shown that this results in lowered bromate formation.
4. Although the hydrodynamic model used in this work resulted in a good agreement with the available data, the use of adjusted parameters should be reduced to the minimum for the design of ozone transfer. Parameters for modeling bromate formation need to be included in this model.
5. The *Ozone System Energy Optimization Handbook* (DeMers et al., 1996) includes only an investigation of FBD contactors. The objective of that study was to develop and demonstrate energy savings opportunities while achieving water quality goals at water treatment plants using ozone. Use of a static mixer system is another step in the optimization of energy for ozone applications by increasing ozone transfer efficiency, optimizing Ct and thereby reducing capital and O & M costs. These benefits need to be incorporated in the *Handbook* and in AwwaRF's Integrated Disinfection Design Framework (IDDF) Project. This should provide a good scientific basis for utilities considering to implement ozonation at their treatment plants.

CHAPTER 1
INTRODUCTION AND BACKGROUND

INTRODUCTION

Since 1989, US EPA Surface Water Treatment Rule (SWTR) requirement for 3 log removal or inactivation of the pathogenic protozoan *Giardia lamblia* has been driving the performance requirements for water treatment processes. Cysts of *Giardia lamblia* are more difficult to inactivate than vegetative bacteria or viruses, requiring higher ozone residuals, and/or longer contact times for equal levels of kill. The current microorganism of interest in drinking water treatment is another pathogenic protozoan, *Cryptosporidium parvum. Cryptosporidium parvum* oocysts have proven to be more difficult to physically remove and inactivate by chemical disinfection than *Giardia lamblia* cysts. The recently promulgated Interim Enhanced Surface Water Treatment Rule (IESWTR) requires 2 Log removal of *Cryptosporidium parvum*. The proposed Long Term 2 Enhanced Surface Water Treatment Rule (LT2ESWTR) places additional treatment requirements (up to 2.5 Log reduction) for WTPs employing conventional treatment that are in full compliance with the IESWTR, based on source water characterization of average oocyst occurrence (Scharfenaker, 2000). These regulatory trends tend to place more stringent performance demands on disinfection systems and have, therefore, increased the need for improvements in design.

Interest in ozonation in water treatment within United States has increased substantially in the 1990s because of suspected carcinogenic properties of trihalomethanes that are formed when organic compounds, naturally present in water, react with chlorine. Among the known disinfectants, ozone has been demonstrated to be very effective in inactivating protozoan cysts and oocysts. However, application of ozone poses several obstacles in water treatment. Gaseous ozone is less readily dissolved in water, tends to be readily consumed by reaction with contaminants other than microbes when in solution, decays more rapidly, making it difficult to establish and maintain a residual. These characteristics create unique challenges for the design of ozone disinfection processes. The static mixer offers one alternative method for improving the efficiency of the dissolution of ozone gas and for optimizing the use of ozone in the disinfection process. Compared to more traditional contacting devices, such as mechanically stirred tanks, turbine agitators, and bubble column contactors, static mixers offer the following advantages: 1) the same amount of ozone transfer can be achieved in a vessel of much smaller size, shorter hydraulic retention time, and with good transfer efficiency, 2) near plug flow with a high degree of radial mixing and very little backmixing produces a very uniform dissolved ozone residual and prevents hydraulic short circuiting (this ensures that all microorganisms in the water are exposed to uniform ozone concentration), 3) low maintenance, reliability and retrofitability, 4) there is potential for enhancement of microbial inactivation by virtue of combination of hydrodynamic shear forces within the mixer and intimate contact with dissolved, and possibly gas phase ozone, and 5) incorporation of a static mixer into a disinfection system will enable better approach to plug flow mixing in the downstream residence time contactors, and will thereby enhance the microbial inactivation and Ct credit of the overall process.

This research project aimed to demonstrate that the use of static mixers can significantly improve the ozonation process in terms of both increased ozone transfer efficiency and enhanced chemical reactions when compared to traditional bubble contactors. Previously observed but not fully explored reaction enhancement takes the form of effective microbial inactivation and pesticide (e.g., atrazine) destruction within seconds in the static mixer unit instead of 5 to 10 minutes for classical diffuser chambers. If confirmed, the microbial inactivation and the increased removal of contaminants in much shorter hydraulic retention times achieved through static mixers could significantly reduce the cost of ozone contactors. This would make ozonation more economical and allow easier retrofits to plants in need for upgrading to meet EPA's rules on surface water treatment or disinfection byproducts.

An in-depth comprehensive approach comprising of both laboratory and pilot-scale testing, supported by data from full scale plants was proposed. Laboratory-scale tests would establish inactivation of regulated microbial contaminants and surrogates over a range of pH and temperature conditions. Pilot-tests at various sites would assess inactivation of microbial surrogates, destruction of pesticide, and formation of bromate over a range of natural water quality and spiked conditions. Pilot tests would allow side-by-side process performance comparisons of static mixer systems and fine bubble diffusion (FBD) contactor systems, as well as provide data for modeling reactor hydrodynamics and process performance of static mixers. Data from full-scale plants fitted with static mixers for ozonation would be used to confirm findings of the laboratory and pilot–scale testing as well as to verify hydraulic and kinetic models. The focus of the investigations would be inactivation of microbial contaminants and surrogates, destruction of pesticides and micropollutants, and occurrence of by-products such as bromate formation.

The project sponsors were AwwaRF and Vivendi Water (VW). Also contributing was Vivendi Water Partnership (VWP), in the UK. VW conducted the pilot-scale study and the laboratory-scale study was done by the University of Alberta. Hydrodynamic and mass transfer modeling was done by the University of Rennes in France and bromate formation modeling by the University of Colorado. Participating utilities were Los Angeles Department of Water and Power, CA; Elizabethtown Water Company, NJ; Cleveland Division of Water, OH; Gwinnett County, GA; Lake Manatee Water Treatment Plant, FL; and VWP in U.K., all of which hosted pilot demonstrations and presented a variety of different surface waters. For full scale demonstrations, VWP in the United Kingdom were utility contributors, and additional pilot work was sponsored by Suburb of Paris, France (VW).

OBJECTIVES

The overall objective of this research was to quantify the potential benefits of ozone application through the use of static mixers in terms of increased transfer efficiency, disinfection capacity, and enhanced chemical reactions at laboratory-scale, pilot-scale and full-scale systems. Specific goals were to identify and quantify the effect of several water quality parameters (e.g., pH, DOC, alkalinity, etc) and environmental/engineered factors (e.g., temperature, transferred ozone dose, and liquid velocity) on the disinfection capacity of the ozone-static mixer system and to assess bromate formation under optimum conditions for microbial inactivation.

LITERATURE REVIEW

Influence of Gas-Liquid Contacting on Microorganism Inactivation by Ozone

It has been proposed that contact with bubbles of gaseous ozone may enhance microorganism inactivation in ozone contactors (Haas 1990; Masschelein et al. 1975; Masschelein et al. 1976; Masschelein 1982b). Gaseous ozone has been shown to be an effective bacteriocide as long as sufficient humidity was present (Dahi and Lund 1980; Ishizaki et al. 1986). The presence of ozone gas bubbles together with dissolved ozone was found to enhance the inactivation of the bacteria *Mycobacterium fortuitum* and the yeast *Candida parapsilosis* by about 1 log-unit as compared to exposure to dissolved ozone alone (Farooq et al. 1977). The authors speculated that, because of their surface active properties, microorganisms would tend to concentrate in the liquid film at the gas-liquid interface, and would thus be exposed to both gaseous ozone and a higher concentration of liquid ozone than in the bulk liquid (Farooq et al. 1978). In a continuous-flow laboratory-scale gas/liquid separator, standard plate count bacteria in secondary wastewater were more effectively inactivated when the ozone bubbles were smaller, given the same dissolved ozone concentration (Ahmad and Farooq 1984). It was postulated that this was due to the increased gas-liquid contact surface area associated with the finer gas bubbles. In a comparison of different contacting systems, the ozone dose necessary for a given level of *Actinomycetes* was reported to be lower if the energy dissipated during gas-liquid mixing was higher (Masschelein 1982b). The same authors also reported that the rate of ozone reaction with picric acid, a slowly reacting phenolic compound, was directly related to the surface area of gas-liquid contact in a dispersed bubble reactor.

The findings of other researchers, however, tend to contradict the gas-bubble hypothesis. Scaccia and Rosen (1978) argued that inactivation of microorganisms would be equivalent in any type of contactor as long as the absorbed ozone dose was the same. They provided supporting evidence for their argument by comparing fecal coliform inactivation in three types of gas-liquid ozone contactors. Finch and Smith (1991) found that inactivation of *E. coli* in wastewater exposed to gaseous ozone in a semi-batch gas-liquid contact tank stirred with a Rushton turbine was less efficient at higher rates of energy input. They concluded that the design of ozone contactors for microorganism reduction should ideally consist of a gas-liquid contact stage for mass transfer, followed by a plug flow reactor for dissolved ozone contact and microorganism reduction.

The varied results of these experiments suggest that the role of ozone bubbles and gas-liquid contact may be system dependent. In addition, the studies were limited to microorganisms that are relatively rapidly inactivated by ozone, such as vegetative bacteria or yeast. No information is available on the effects of gas-liquid contact on the comparatively ozone resistant microorganisms such as sporulated bacteria or encysted protozoa like *Cryptosporidium parvum* or *Giardia lamblia*.

Use of Static Mixers for Improving Microorganism Inactivation

Static mixers, also called motion-less mixers, consist of a series of stationary mixing elements mounted inside a section of pipe or duct. They have been used in chemical processing and the food industry for applications ranging from the mixing of highly viscous fluids, like polymers, to absorbance of gases into non-viscous liquids (Myers et al. 1997; Rader et al. 1989). As the fluid mixture flows through the elements of the static mixer, it is sheared, sub-divided into smaller elements, and recombined many times. Fluid kinetic energy is rapidly dissipated resulting in a highly micro-mixed, homogenous fluid with a radially uniform concentration. In gas-liquid contacting systems, high rates of mass transfer are promoted by formation of very fine bubbles, the surfaces of which are renewed constantly by coalescence and redistribution of the fluids, and by high turbulence (Grosz-Röll et al. 1982). The short residence times and rapid energy dissipation within static mixers make them particularly useful for enhancing the yield of very fast and competitive reactions where mass transfer and the degree of micro-mixing are the rate controlling steps (Bourne et al. 1992). For this reason, static mixers are useful for the rapid mixing of coagulation chemicals in water treatment (Schulgen et al. 1996, Amirtharajah et al. 2001).

Static mixers, have been available as an option for ozone dissolution for some time (Masschelein 1982a). Langlais and co-authors (1991) list the principal advantages of static mixers, and other in-line contactors, for ozone contacting as:

1. they are simple systems with no moving parts that require little maintenance;
2. they provide good mixing and mass transfer; and
3. because of the very short residence time much smaller contacting vessels are required compared to FBD contactors.

The main disadvantage of static mixers in a drinking water treatment facilitiy is that the pressure drop incurred may create the need for pumping and additional energy expenditure. The benefits of using static mixers, or any other in-line mixing devices, for improving the efficiency of microorganism reduction processes in water treatment have been explored only to a limited extent. Zhu and co-workers (1989a) found that, in a laboratory static mixer system, inactivation of *E. coli* in wastewater was controlled by the mass transfer rate within the static mixer and even proposed a mathematical relationship between inactivation and the mass transfer coefficient. However, it is not clear from the laboratory design if these researchers were able to separate the effects on the microorganisms of mass transfer from those of contact with dissolved ozone. After studying the hydrodynamics of several conventional FBD contactors, Martin and co-workers (1992) recommended the use of static mixers as an alternative for ozone dissolution. They argued that a static mixer used in combination with a plug-flow contactor would provide the best microorganism reduction. This design concept of a mass transfer stage followed by microorganism reduction stage is very similar to that suggested by Finch and Smith (1991). The concept was later tested in a pilot study, however, the focus of the work was on ozone transfer efficiency and oxidation of atrazine. Very limited information on the reduction of heterotrophic plate count (HPC) bacteria was provided (Martin and Galey 1994).

Others studied HPC and fecal coliform bacteria reduction in a static mixer ozone contactor (Bonnard et al. 1999). The laboratory design, however, did not allow the effect of the static mixer microorganism reduction to be determined separately from other system parameters. Recently, enhanced reduction of *Cryptosporidium parvum* oocysts was reported when the microorganisms were pre-contacted with dissolved chlorine in a static mixer relative to pre-contact in an empty pipe (Heindel et al. 1999; Amirtharajah et al. 2001). This finding suggests that vigorous mixing within the static mixer might increase the efficacy of the chlorine by loosening, or otherwise altering, the oocyst wall.

Few documented full-scale examples of static mixer ozonation systems are available. Zhu (1990) described a small installation in which hospital wastewater was contacted with ozone in a static mixer for a 0.1 min contact time with applied ozone doses ranging from 20 to 45 mg/L.

Reduction of *E. coli* in the wastewater was 3 to 5 log-units. Martin and co-workers (1995) described an 84 ML/d static mixer installation at the Melun-Arvigny water treatment plant in France. Ozone was contacted with a side-stream of plant water flow in one static mixer and was then further blended with the main flow in a second static mixer. Additional contact time of 6.2 minutes with dissolved ozone was provided in an "up and over" style, 4-chamber contactor. No microorganism reduction or hydrodynamic performance data were available.

Use of Bacterial Spores as Models for Ozone Inactivation of Encysted Protozoa

Traditional indicators of microbial water quality, like the total and fecal coliform bacteria, are not appropriate for determining the inactivation effectiveness of protozoan parasites by ozone. Proposed alternative surrogate microorganisms for assessment of treatment processes include sporulated aerobic bacteria (Nieminski and Bellamy 1998; Rice et al. 1996; Toenniessen and Johnson 1970), sporulated sulphite-reducing anaerobes (Hijnen et al. 1997) and heterotrophic plate count bacteria (Nieminski and Bellamy 1998). The properties of an ideal model organism that can be used to evaluate ozonation processes include:

- comparable resistance to ozone as *Cryptosporidium parvum,* with the same sensitivity to water quality variables like temperature and pH;
- predictable ozone inactivation kinetics;
- easy and inexpensive to produce in substantial titres;
- easy and inexpensive to enumerate; and
- non-pathogenic to humans.

Aerobic bacterial spores of *Bacillus* species have been shown to demonstrate considerable resistance to inactivation by ozone (Botzenhart et al. 1993; Broadwater et al. 1973; Choe 1998; Facile et al. 2000, Owens et al. 2000; and Herbold et al. 1989). In batch reactors, cultured spores of *Bacillus cereus* were found to have similar resistance to ozone as *Cryptosporidium parvum* oocysts (Finch and Choe 1999). However, *Bacillus cereus* is an agent of food poisoning in humans (Goepfert et al. 1972) and is, therefore, not an ideal model organism. In pilot-trials, both indigeneous aerobic spores and a pure strain of *Bacillus subtilis* spores were found to be more than twice as resistant to ozone as were *Cryptosporidium parvum* oocysts (Owens et al. 2000) suggesting that these non-human pathogenic species of *Bacillus* might serve as good model organisms for ozonation processes.

The kinetics of inactivation of natural aerobic spores and a pure strain of *Bacillus subtilis* spores by ozone were measured at 22°C and were modeled using a modified Hom equation (Facile et al. 2000). At the 2 log-unit inactivation level, resistance of the spores to ozone was found to be comparable to that reported for *Cryptosporidium parvum* oocysts in other work. However, unlike *Cryptosporidium parvum*, the kinetics of spore inactivation were found to be a significant function of pH. Further, indigenous spores were found to be less resistant to ozone than were the cultured spores. The authors cautioned that the resistance to ozone may depend on the source of the spores. Others have also reported that the resistance of various species of aerobic spores to chemical agents may differ (Broadwater et al. 1973, Barbeau et al. 1999).

Previous Studies

Design of effective ozone disinfection systems using static mixers will require a solid understanding of how the operating variables of the static mixer influence the inactivation of microbial contaminants. Very little research work has been reported in which the effectiveness of the static mixer as an ozone contacting device for drinking water disinfection has been demonstrated. Published information on disinfection with ozone using a static mixer as a contacting device is limited to bacteria and viruses which are relatively easy to inactivate with ozone (Martin and Galey 1994; Zhu 1991; Zhu et al. 1989a). The Water Industry Database lists approximately 65 plants in the U.S. which use static mixers for the rapid mixing of coagulants (AWWA and AwwaRF 1992). To the best of our knowledge, there is no published information on the inactivation of protozoan cysts, such as *Giardia lamblia*, and *Cryptosporidium parvum* using ozone and a static mixer.

The static mixer project also offered an opportunity to demonstrate the use of kinetic models of microbial inactivation developed in batch experiments for flow systems, and to further develop rational approaches to disinfection design and engineering analysis. In recent years, much research effort has been dedicated to the development of rigorous kinetic models for the inactivation of *Giardia lamblia* and *Cryptosporidium parvum* (Finch et al. 1994; Finch et al. 1993b; Finch et al. 1997; Gyürék et al. 1997; Gyürék et al. 1998; Haas et al. 1996; Haas and Joffe 1994). These models have been developed in batch disinfection experiments, and mostly using laboratory waters. Advanced kinetic models, such as the Incomplete Gamma Hom (IgH) with disinfectant decay, have been adopted to describe the non-linear characteristics of microbial inactivation that have been observed experimentally. These models can also account for the decay of unstable disinfectant chemicals such as ozone. What remains to be demonstrated is whether these models can provide reliable predictions of microbial inactivation in non-ideally mixed flow systems.

The US EPA SWTR has promoted the use of the concentration-time product, the Ct product, as the basis for engineering analysis and design of disinfection processes (Malcolm Pirnie Inc. and HDR Engineering Inc. 1991). Although this approach is conceptually simple, it does not address the complexities of full-scale ozone disinfection process, which include non-linear inactivation kinetics, mass transfer limitations, disinfectant decay and non-ideal flow patterns. Several authors have pointed out the weaknesses of the Ct approach and have proposed a more rational alternative that is based on chemical reaction engineering analysis (Lawler and Singer 1993; Lev and Regli 1992a; Lev and Regli 1992b; Roustan et al. 1993). This approach involves integration of batch kinetics and disinfectant decay, with the hydraulic and mass transfer characteristics of the disinfection contactor. Only very limited research work has been reported in which batch inactivation kinetics have been integrated with hydraulic characteristics of flow systems (Haas et al. 1994; Tomiak et al. 1998). The reason for the sparseness of research of this type may be that there are practical difficulties in carrying out careful disinfection work on flow systems, especially at pilot- or full-scale. These difficulties are particularly acute for the protozoans, *Giardia lamblia* and *Cryptosporidium parvum.*

In summary, limited information exists on microbial inactivation by ozone through a static mixer. This study is expected to improve our understanding of the disinfection process by ozone application through a static mixer at lab-, pilot- and full-scale levels. This study will also assist us in determining if kinetic models of microbial inactivation determined in batch experiments can be used to produce reliable predictions of inactivation in the static mixer flow system.

Table 1.1
Task and associated study with the task

Task	Study	Chapter
Task I	Laboratory-scale studies	2
Task II	Pilot-scale studies	3
Task III	Full-scale studies	4
Task IV	Hydrodynamic Modeling Bromate Modeling	5 and 6

PROJECT DESCRIPTION

The project was comprised of four major tasks. Those tasks and the study associated with these tasks are shown in Table 1.1. Also shown in Table 1.1 is the Chapter in which the information pertaining to a particular study can be found in this report.

The combination of the three hierarchies of testing served to address concerns raised during prior laboratory studies about scale-up from laboratory scale to pilot-scale or full-scale. The synthesis of this work was a modeling framework encompassing the hydrodynamics and reaction chemistry of static mixers. Bromate formation was also modeled. *Cryprosporidium parvum* inactivation was modeled as part of the bench-scale study.

Experimental Plan

To address the issue of scale-up, to ensure that the bench- and pilot-scale studies be designed to best represent full-scale treatment plants, and the results from three phases complement one another, the following experimental plan was developed:

1. Two testing units, one for lab-scale (flow = 7.6 to 22.7 L/min, 2 to 6 gpm) and one for pilot-scale (flow = 3.8 to 38 L/min, 1 to 10 gpm) were built simultaneously at the University of Alberta to conduct side-by-side studies and address the issue of scale-up. Laboratory- and pilot-scale studies were done with a 15.7 mm (0.6 inch) diameter static mixer. A major portion of the work was done with *Bacillus subtilis* in the bench and pilot-scale studies. Few trials with *Giardia muris* and *Cryptosporidium parvum* were conducted in the later phases of the lab-scale study to test the design predictions for the apparatus and to determine the effect of ozone gas/liquid contact in a static mixer on inactivating these microorganisms.
2. To have meaningful comparisons between pilot-scale and full-scale, side-by-side testing was conducted in the U.K. at the Chertsey WTP with the pilot unit operated next to a full-scale static mixer. Native aerobic spore-formers and *Bacillus subtilis* were also monitored at full-scale. The inactivation of *Bacillus subtilis* by ozone formed the basis for comparing the three scales of process examined in this project.
3. Full-scale testing in the U.K. studied microbial inactivation, pesticide/micropollutant destruction and bromate formation under typical operating conditions at four water treatment plants (WTPs) fitted with static mixers for ozonation. Data collected from the three testing phases was processed and used in modeling.

CHAPTER 2
LABORATORY-SCALE STUDY

OVERVIEW AND OBJECTIVES

The objective of the laboratory study was to determine how a static mixer can be integrated with an ozone contacting system to maximize microorganism inactivation. The laboratory study included investigations of microorganism inactivation in both batch reactors and in a small-scale laboratory static mixer ozone contactor. Microorganisms that were studied included a cultured strain of *Bacillus subtilis* spores, *Cryptosporidium parvum* oocysts and *Giardia muris* cysts. The specific goals were to:

1. determine the inactivation kinetics of the *Bacillus subtilis* spores by ozone over a range of temperature and pH values, and determine the suitability of these spores as surrogates for *Cryptosporidium parvum* oocysts,
2. determine the efficiency of microorganism inactivation in the static mixer contactor by comparison to inactivation measured in batch reactors,
3. measure direct microorganism reduction within the static mixer,
4. determine the relationship between the hydrodynamic conditions in the static mixer, ozone transfer efficiency, and inactivation of microorganisms.

These goals were achieved by conducting challenge experiments in both batch and static mixer ozone contacting systems under controlled laboratory conditions.

MATERIALS AND METHODS

To address the issue of scale-up, to ensure that the bench- and pilot-scale studies be designed to best represent full-scale treatment plants, and the results from three phases complement one another, two testing units, one for lab-scale (flow = 7.6 to 22.7 L/min, 2 to 6 gpm) and one for pilot-scale (flow = 3.8 to 38 L/min, 1 to 10 gpm) were built at the University of Alberta. The "front ends" of the two units were essentially the same. The "front end" consisted of a water and microbe feed system, a 15.7 mm (0.6 inch) ID Sulzer SMV for dissolution of ozone and a 100 mm (4 inch) ID gas-liquid separator column with a 60 second contact time. The main difference between the two units was in the scale of the downstream disinfection contactor sections. The pilot unit contactor was about 290 L volume and provided 25 minutes of contact time at 11.4 L/min (3 gpm). The lab unit contactor was 29 L and provided 25 minutes of contact time at 1.14 L/min (0.3 gpm). The physical configurations were slightly different as well. The lab unit was a serpentine type contactor with 6 vertical sections while the pilot unit was 4 columns connected in series. Additional details of the pilot unit are provided in Chapter 3. Both units were designed to provide close to a plug flow mixing condition in the contactors. Plug flow was confirmed by tracer test results that are presented in this Chapter and in Chapter 3. There was, therefore, no scale-up issue between the static mixer component of the two laboratory static mixer systems because they were the same. The scale-up issues were related to the disinfection sections where the microbes to be disinfected were in contact with the dissolved ozone. To address this

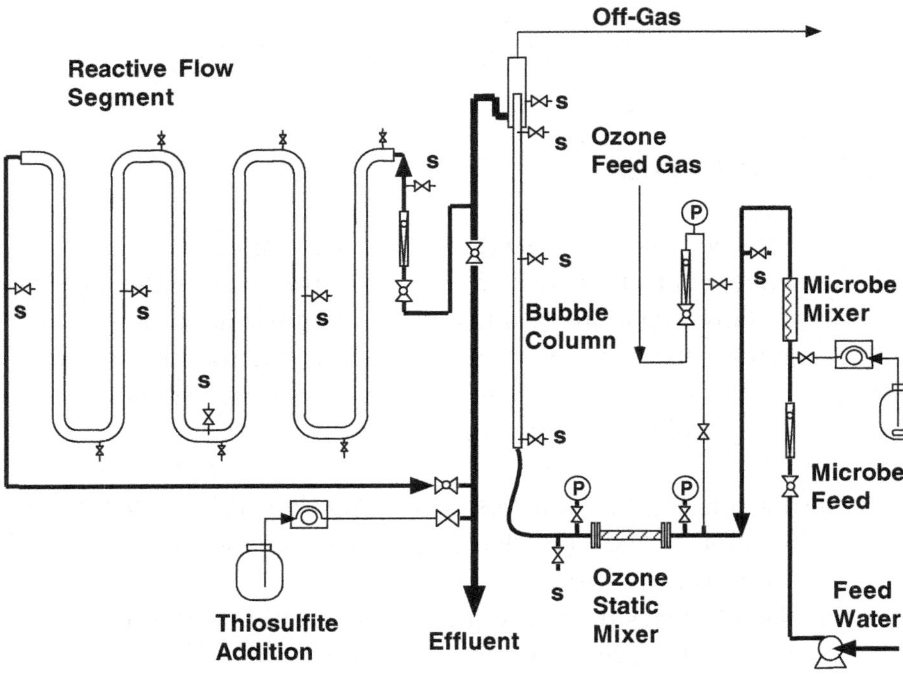

Note: The 's' symbols denote a liquid sampling point.

Figure 2.1 Detailed schematic of the laboratory static mixer ozone contactor

issue side-by side tests with both units were conducted and the results are presented in this chapter.

Description of the Laboratory Static Mixer Contactor

A laboratory-scale prototype of a static mixer contactor was built and operated at the University of Alberta in Edmonton, Alberta. The contactor consisted of a static mixer for initial gas-liquid contacting and dissolution, a short gas/liquid separator for additional gas-liquid contacting and off-gas separation (gas/liquid separator), and a reactive flow segment for extended contact between microorganisms and dissolved ozone. A detailed schematic diagram of the laboratory static mixer ozone contactor is provided in Figure 2.1. A photograph of the same unit is shown in Figure 2.2. For all experiments, the water source was building tap water supplied from the Epcor Rossdale Water Treatment Plant in Edmonton. This plant provides full conventional treatment including coagulation with alum and polymers, flocculation, settling, lime softening, chlorination, dual media filtration and ammonia addition. Hot and cold tap water sources were blended to provide water at the desired laboratory temperature to within ± 1°C. The blended water was passed through a set of four parallel activated carbon (F-400, Calgon Carbon, Pittsburgh, PA) columns to reduce the combined chlorine concentration from 2 mg/L to less than 0.2 mg/L. The de-chlorinated water was then diverted to a 1000 L polyethylene storage tank and was pumped to the inlet of the laboratory static mixer contactor.

Ozonized gas was produced by passing extra-dry oxygen (Praxair, Edmonton, AB) through an air-cooled corona discharge generator (Doman Ozone, Edmonton, AB). A regulated flow of ozonized gas flow was introduced into the center of a 1.3 mm ID polyvinyl chloride

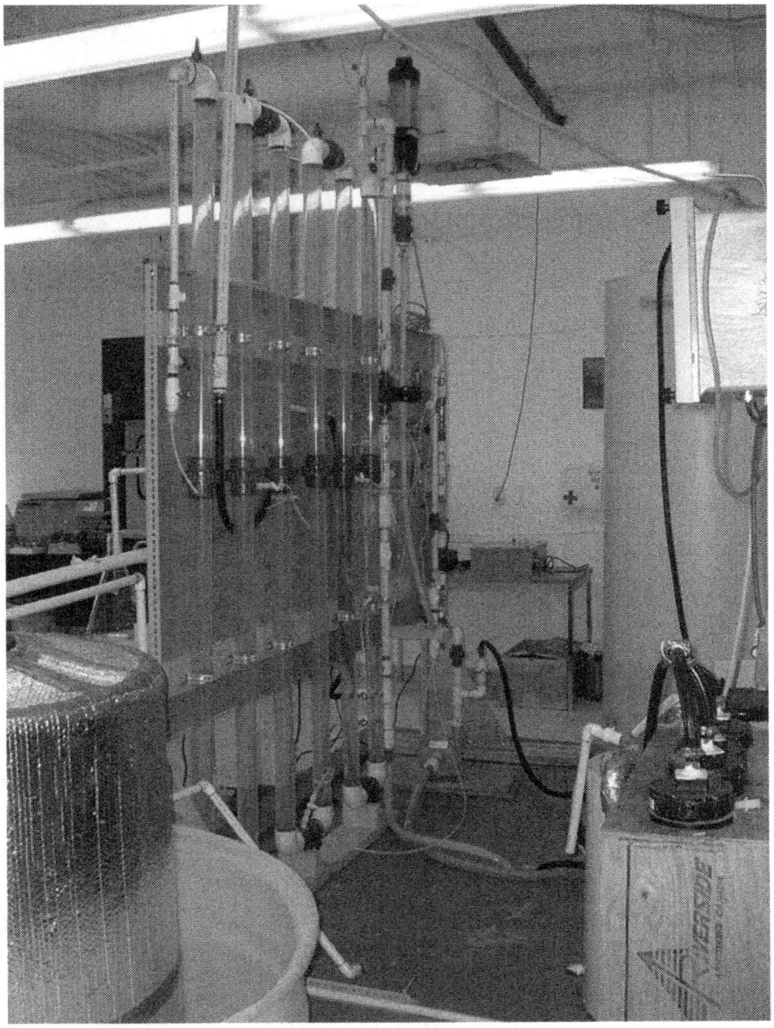

Figure 2.2 Photograph of the laboratory static mixer ozone contactor

(PVC) pipe directly upstream of the static mixer through a 3 mm Teflon® tube. The water-gas mixture then flowed through a 15.7 mm ID Sulzer SMV static mixer fitted with three 15 mm corrugated-type structured stainless-steel elements. The arrangement of the static mixer elements within the static mixer spool piece is depicted in Figure 2.3. Including the space between the elements, the total internal mixing length and volume were 75 mm and 15 mL, respectively. A similar arrangement was used for the static mixer employed in the pilot studies. The gas-liquid mixture leaving the static mixer flowed into the bottom of a 38 mm ID vertical gas/liquid separator via a 16 mm ID × 1 m section of clear polyvinyl chloride tubing. Off-gas was separated from the liquid at the top of the column by means of a circular overflow weir. The height of the column could be adjusted to 0.5, 1.2 or 2.3 m by adding or removing column sections.

Flow of ozonated water from the weir was split. About 2 L/min was diverted to the reactive flow segment and the remainder was diverted to waste. This was done to minimize the size (and cost) of the contact columns. This feature also allowed the conditions of the static mixer and gas/liquid separator to be de-coupled from the conditions in the disinfection contactors. That is, liquid velocity in the static mixer was varied independently of contact time in the disinfection

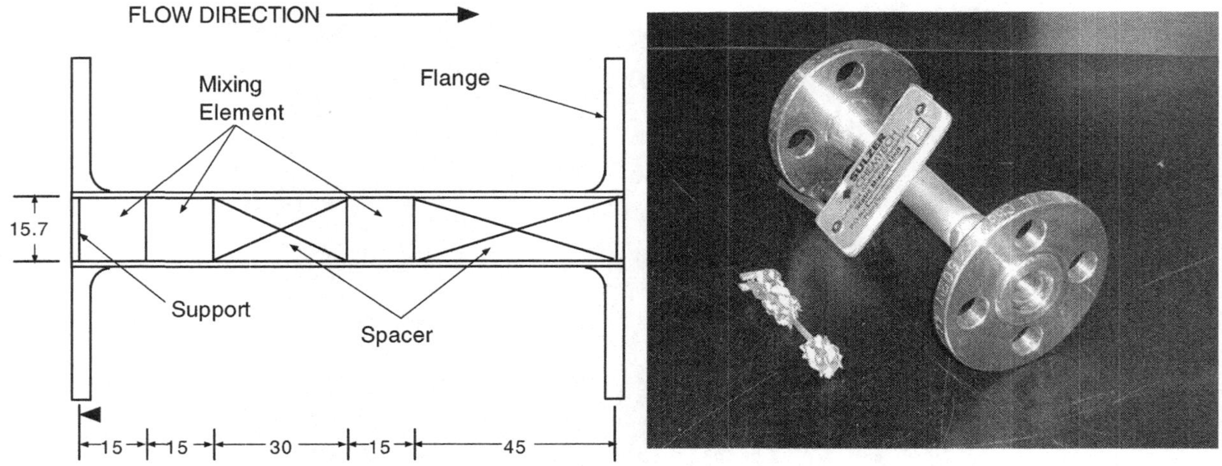

Note: A similar static mixer was used for the pilot unit. All dimensions are in mm.

Figure 2.3 Detail of the internals of the Sulzer SMV static mixer employed for the laboratory unit (left) and photograph of static mixer and mixing elements (right)

contactors. The reactive flow segment consisted of a 15 m length of 51 mm ID PVC, arranged in a serpentine fashion with six vertical sections of 2.2 m length each and six horizontal sections of 0.28 m each connected with 90° elbow bends. The length of the reactive flow segment could be varied between laboratory trials by adding or removing piping sections. The serpentine arrangement of the reactive flow segment was merely for convenience.

For microorganism challenge experiments, a well-mixed suspension of microorganisms was continuously introduced into the feed water at a controlled rate using a variable speed peristaltic pump. The microorganisms were blended with the flowing water by means of an in-line mixer located downstream of the microorganism addition point, but upstream of the ozone addition point and ozone static mixer. Capability for withdrawal of liquid samples was provided at various locations in the system as indicated in Figure 2.1. All sampling tubes and valves were constructed of Teflon® to minimize reaction with dissolved ozone, and were designed to withdraw the sample from the flowing liquid at a point located at a distance from the inside of pipe equivalent to 1/3 of the pipe diameter. Provision was also made for sampling of the ozonized feed gas and the off-gas from the top of the gas/liquid separator. Water flows were manually controlled and were measured by calibrated, in-line flow meters. The pressure at the inlet and the outlet of the static mixer and at the outlet of feed gas rotameter was measured by pressure gauges. The feed gas rotameter was calibrated using a wet test meter (GCA/Precision Scientific, Chicago, IL). Gas flow was corrected for system pressure and temperature and reported in normal litres per minute (L/min) at 0°C and 760 mm Hg.

Residual ozone in the effluent streams from the reactive flow segment and the gas/liquid separator was neutralized by continuous addition of sodium thiosulfite solution. For experiments with the human pathogenic organism *Cryptosporidium parvum*, all contaminated effluent was collected into a containment tank and decontaminated by heating to 70°C. Most of the system components were constructed of polyvinyl chloride (PVC) plastic to minimize reaction with ozone. The main sections of the gas/liquid separator and the reactive flow segment were fabricated of clear PVC to enable visual observation of the mixing pattern during tracer experiments with methylene blue dye.

Table 2.1
Manipulated variables in the laboratory static mixer ozone contactor

Variable	Symbol	Unit	Nominal Range
Feed Water Flow Rate	L	L/min	7.5 to 15
Static Mixer Superficial Velocity	v_s	m/s	0.7 to 1.4
Gas to Liquid Volumetric Flow Rate Ratio @ 0°C and 1 atm	G/L	%	1.2 to 4.1
Gas/Liquid Separator Height	H	mm	45, 121, 225
Initial Ozone Concentration (at Reactive Flow Segment Inlet)	C_0	mg/L	0.1 to 1.2
Cumulative Theoretical Hydraulic Residence Time	t	min	4.5, 6, 12.2, 16

Operating Ranges and Variables

The laboratory static mixer contactor was designed to provide rapid and efficient ozone dissolution within the static mixer and gas/liquid separator and efficient microorganism reduction in the reactive flow segment. Duration of contact within the mixing elements of the static mixer was less than 0.2 s and total gas-liquid contact within the static mixer and gas/liquid separator was usually less than 45 s. The bulk of the contact between dissolved ozone and the microorganisms took place under much more gentle flow conditions in the reactive flow segment. The high *L/D* ratio of the reactive flow segment (up to 300:1 at maximum length of 15 m) minimized the potential for back-mixing and flow short-circuiting.

The principle manipulated variables in the laboratory static mixer ozone contactor and the nominal ranges of each are listed in Table 2.1. An important laboratory feature of the system was the ability to independently control the flow rates through the static mixer and through the reactive flow segment. In this way, the hydrodynamic conditions within the static mixer, given by the superficial velocity, v_s, and the gas-liquid volumetric flow rate ratio, *G/L*, could be manipulated independently of the conditions of dissolved ozone contact in the reactive flow segment. The latter were determined by the initial ozone concentration in the reactive flow segment, C_0, and the hydraulic residence time, HRT. For a given combination of water flow, *L*, and *G/L*, a desired C_0 target was arrived at by adjusting the concentration of ozone in the supplied ozone feed gas. The hydraulic residence time, τ, was determined by the flow rate of water through the reactive flow segment and the length (and, therefore, volume) of the reactive flow segment. As indicated in Table 2.1, the length of the reactive flow segment and the total contact time was varied between experiments by adding or removing sections of piping.

Transfer Efficiency Measurements

The ozone transfer efficiency, *TE,* was defined according to the following equation (Langlais et al. 1991):

$$TE\,(\%) = \frac{Y_f - Y_o}{Y_f} \times 100\% \tag{2.1}$$

where Y_f and Y_o are the concentration of ozone in the feed-gas and off-gas (in mg/L), respectively. For the laboratory static mixer ozone contactor, Y_f was the concentration of ozone in the gas from the ozone generator and Y_o was the concentration of ozone in the off-gas leaving the top of the gas/liquid separator. This transfer efficiency measurement, therefore, incorporated the ozone transfer that occured in the static mixer, the gas/liquid separator and any interconnecting piping and tubing. The ozone concentrations in the feed-gas and in the off-gas were measured by the potassium iodide absorption method recommended by the International Ozone Association for determining gas-phase ozone concentrations (Masschelein et al. 1998).

With the static mixer unit operating at steady-state, and after sufficient time was allowed to purge the gas space above the liquid at the top of the gas/liquid separator, the gas flow was diverted to a gas washing bottle filled with 300 mL of a solution of acidified 5% reagent grade potassium iodide prepared in de-ionized water. The sparging tube of the washing bottle was fitted with a drawn tip with a 0.5 mm diameter orifice. The total volume of gas processed through the wash bottle was measured with a wet test meter (GCA/Precision Scientific), corrected for ambient temperature, pressure and moisture content and reported in normal litres (L) of dry, ozone-free gas at 0°C, 1 atm. The gas sample size for each analysis was between 1 and 2 L. The ozone concentration in the gas was calculated from the following equation:

$$Y = 24\frac{N_{thio}V_{thio}}{V_g} \tag{2.2}$$

where N_{thio} was the normality of the thiosulfate titrant (0.1 N), V_{thio} was the volume of the thiosulfate titrant added (mL), and V_g was the total volume of the ozone gas collected in litres (L) at 0°C and 760 mmHg.

Residence Time Distribution

The residence time distribution (RTD) was determined at each microorganism sampling location in the static mixer ozone contactor from the results of pulse-input tracer tests conducted using methylene blue dye. These tests were done for both the laboratory- and pilot-scale units. During the tracer experiments, the power to the ozone generator was shut off to avoid de-colorization of the dye by reaction with ozone. For each experiment, the dye concentration was monitored at the selected downstream location by diverting a side-stream of the flow through a 10 mm quartz spectrophotometric flow cell mounted inside a Hewlett Packard (Wilmington, DE) 8452A diode-array spectrophotometer. The absorbance at 652 nm was measured and recorded at 1-s intervals. The normalized exit age distribution function, *E(t),* at the downstream sampling location was estimated at each sampling instance according to:

$$E(t) = \frac{C(t)}{\int_0^\infty C dt} \cong \frac{C_{avg,t}}{\sum_0^{t_f} C_{avg,t}\Delta t} \tag{2.3}$$

where C_t was the measured tracer absorbance at time t, Δt was the sampling interval of 1 s and $C_{avg,t}$ was the average dye absorbance at time t given by $(C_t\text{-}C_{t\text{-}1})/2$. Tracer experiments were conducted in triplicate for each sample point.

Challenge Experiments With Microorganisms

For all experiments involving microorganisms, the static mixer contactor was operated at steady-state, with continuous addition of water, ozonized gas and microorganism suspension. Prior to the initiation of microorganism addition, the static mixer contactor was operated at the target steady-state ozonation condition for approximately 30 min to allow the ozone concentration in the water to stabilize. At time = 0, metering of the prepared microorganism suspension into the contactor feed was commenced at a rate of 0.1 to 0.3 L/min. The suspension concentration and flow were chosen to provide the following nominal microorganism concentrations in the static mixer feed stream:

Bacillus subtilis spores	0.2 to 1×10^9 /L
Cryptosporidium parvum oocysts	1×10^6 /L
Giardia muris cysts	1×10^5 /L

The time required for 98% of the injected substance to exit the system (t_{98}) which was determined from the measured residence time distribution, was used as the criteria for achieving steady-state operation after the microorganism flow was first introduced. Once the t_{98} was exceeded, samples were collected from various points in the system for both microorganism analysis and determination of dissolved ozone concentration.

Two methods were used for determination of the aqueous-phase ozone concentration; (1) direct ultraviolet (UV) absorbance and (2) the indigo trisulphonate colorimetric method. The UV method was selected as the primary measurement technique to allow for a direct comparison to the available kinetic models for infectivity reduction of *Cryptosporidium parvum* (Gyürék et al. 1999; Li et al. 2001) and *Giardia muris* (Finch et al. 1993a). Samples were collected directly into either 10 mm or 50 mm quartz cuvettes and the absorbance at 260 nm was immediately measured on an Ultrospec 2000 UV/visible spectrophotometer (Pharmacia Biotech, Cambridge, UK). A reference blank was prepared by collecting a 200 mL sample of water from the outlet of the static mixer unit reactive flow segment, after several minutes of contact with dissolved ozone. The ozone in the sample was neutralized by addition of 0.20 mL of 1 M sodium formate solution and the absorbance was measured at 260 nm. The ozone concentration in the sample, *C*, in mg/L was determined from the following equation:

$$C = [A_{260,\ \mathrm{sample}} - A_{260,\ \mathrm{reference}}] \times 14.45 \quad (2.4)$$

where $A_{260,sample}$ was the absorbance of the sample and $A_{260,\ reference}$ was the absorbance of the reference. The proportionality factor 14.45 was derived from the molar absorbance of 3300 $M^{-1}cm^{-1}$ reported by Hart and co-workers (Hart et al. 1983) and the molecular weight of ozone of 48,000 mg mol^{-1}. Parallel measurements of dissolved ozone were made using a modified version of the indigo method of Bader and Hoigné (1981; 1982) with molar absorbance coefficient of 20,000 $M^{-1}cm^{-1}$. Samples for each ozone determination method were collected in triplicate.

Samples for spore enumeration (approximately 100 mL) were collected directly into bottles containing a 0.7 mL of 0.1 N certified sodium thiosulfate standard solution (Fisher Scientific), sufficient to neutralize the remaining ozone residual. These samples were collected in triplicate. Oocysts and cysts were recovered by collecting between 2 and 8 L of the water directly into Erlenmeyer flasks into which an appropriate volume of a 0.1 N thiosulfate standard solution

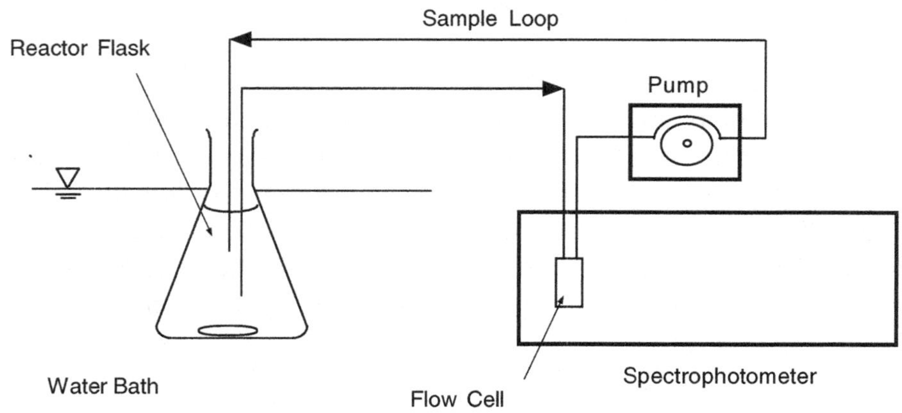

Figure 2.4 Batch ozone reactor laboratory set-up

had been added. For the reactive flow segment, this required collection of the entire effluent flow for a 4 min period. All sample flasks were gently stirred during sampling to ensure immediate neutralization of the dissolved ozone by the sodium thiosulfate. Oocysts and cysts were recovered by filtering the sample through a 47 mm polycarbonate track-etched membrane with a 0.8 micron pore size (Model K08CP04700, Osmonics, Minnetonka, MN) using a suction filtration apparatus. Oocysts were scraped and washed from the membrane using a rubber policeman. For *Giardia muris* experiments, a small amount of 0.01% Tween solution was added to the samples to enhance cyst recovery. Scrapings and wash water were collected into 40 mL Teflon® lined plastic centrifuge tubes for final recovery by centrifugation and for determination of inactivation by infectivity analysis.

Batch Reactor Experiments

To determine the kinetics of *Bacillus subtilis* and *Cryptosporidium parvum* oocyst inactivation by dissolved ozone, experiments were conducted in well-stirred batch reactors using methods and materials similar to those used by others (Gyürék et al. 1999; Li et al. 2001). The batch reactor apparatus is depicted in Figure 2.4. The water matrix for the batch reactor experiments was either samples of water collected from the feed tank of static mixer ozone contactor during challenge experiments, or 0.05 M phosphate buffer at either pH 6 or pH 8. The phosphate buffer was prepared by dissolving appropriate amounts of reagent grade disodium hydrogen orthophosphate and potassium dihydrogen orthophosphate into ultra-pure deionized water from an Elga® system. Buffers were made ozone demand free (ODF) by bubbling ozonized gas through the prepared solution until saturated with dissolved ozone at approximately 20 mg/L, and then letting stand for 1 h. The remaining ozone was removed by bringing the buffer to a boil for 10 minutes. After cooling, the pH was measured and recorded. All material used in the batch reactor experiments, including glassware, stir bars and pipette tips, were made ODF prior to use by soaking in a concentrated (approx. 20 mg/L) dissolved ozone solution followed by air drying.

Suspensions of microorganisms were prepared in 200 mL of water matrix and then exposed to dissolved ozone in 250 mL Erlenmeyer reactor flasks. For experiments with *Bacillus subtilis*, an aliquot of prepared spore stock solution was first diluted in an appropriate volume of the test water in a 50 mL plastic centrifuge tube. The contents of the tube were then mixed vigorously on a vortex mixer for two minutes to reduce spore agglomerates. A 5 mL aliquot of the diluted, vortex-mixed spore suspension was then added to the test water in the reactor flask to provide a starting spore concentration of between 1×10^5 and 2×10^6 CFU/mL. For experiments with *Cryptosporidium parvum*, a suspension of purified and washed oocysts was prepared in about 0.5 mL of de-ionized water in a 5 mL plastic tube. The contents of the tube were vortex mixed at a high setting for 30 seconds and then transferred to the water in the reactor flask. The concentration of oocysts in the flask was 5×10^7 oocysts/mL.

Throughout the contact period, the microorganism suspension was continuously stirred by means of a Teflon®-coated magnetic stir bar. The reactor flask was immersed in a temperature-controlled water bath to maintain the reaction temperature within ± 0.5°C of the specified target. Prior to ozone addition, the microorganism suspension was left in the water bath with stirring for 15 minutes to equilibrate to the laboratory conditions. The temperature of the suspension was measured to ensure that the temperature was within ± 0.5°C of the target. Ozone stock solution was prepared by bubbling ozone gas through 300 mL of water from an Elga® Ultrapure system (Fisher Scientific) water cooled to between 1 and 4°C in ice. An aliquot of ozone stock solution was transferred to the stirred microorganism suspension in the reactor flask using a calibrated Oxford Macroset Pipette fitted with an ODF tip. The tip of the pipette was inserted just below the surface of the suspension in the flask to minimize the loss of ozone by volatilization during the transfer. The concentration of ozone in the stock solution was determined both before and after the transfer by direct measurement of UV absorbance at 260 nm and assuming a molar absorption coefficient of 3,300 $M^{-1}cm^{-1}$. Absorbance measurements were made on a Pharmacia Biotech Ultrospec 2000 UV/visible spectrophotometer using a 10 mm quartz cuvette.

The dissolved ozone concentration in the suspension was monitored continuously by direct UV absorbance on a Hewlett Packard model 8452A diode-array spectrophotometer equipped with a 10 mm quartz continuous-flow cell. A molar absorbance coefficient of 3,300 $M^{-1}cm^{-1}$ (Hart et al. 1983) was assumed. In addition to the UV measurement, samples were removed from the reactor flasks at intervals using an Oxford Macroset Pipette fitted with an ODF tip and analyzed using the indigo trisulphonate colorimetric method. At the end of the prescribed contact time, 0.2 mL of 1 M reagent grade sodium formate solution was added to the flask to neutralize the remaining dissolved ozone. For *Bacillus subtilis* experiments, three 10 mL samples were removed from the flask for enumeration of viable spores by the membrane filtration method. For *Cryptosporidium parvum* experiments, the contents of the reactor were transferred to a 250 mL plastic centrifuge tube and the oocysts recovered by centrifugation. For each set of batch experiments, a control flask was established using the same procedures and conditions as the laboratory flasks, but without addition of ozone.

Bacillus subtilis Methods

Production

A sample of spores of *Bacillus subtilis* strain number 6633 was obtained from American Type Culture Collection (ATCC, Manassas, VA). The freeze-dried spore sample was added to nutrient broth and the broth culture was incubated for 1 to 3 days at 35°C to generate vegetative cells. For propagation of spores, 1 mL of the vegetative cell suspension was inoculated onto a layer of R2A agar in 1 L flasks along with a small amount (15 to 20 mL) of 0.05 M phosphate buffer (pH 6.0). After 14 to 16 days of incubation at 35°C, the spore suspension was collected from the agar and was washed 3 times in ODF pH 8, 0.05 M phosphate buffer (4500 × *g*, 10 min, 4°C). The suspension was then heated to 75°C in a water bath and maintained at this temperature for 20 min to kill any remaining vegetative cells. This spore suspension was then washed one more time in buffer, stored at 4°C and maintained as a stock. To generate working suspensions for experiments, the above process was repeated using a sample of the stock as starting material. Three separately prepared spore stocks were used during the course of this work. Stock A was used in the batch reactor experiments in ozone demand-free buffered water at 3, 12 and 22°C. Stocks B and C were used in the challenge experiments with the static mixer ozone contactor and in the matching set of batch reactor experiments at 16°C.

Enumeration

Viable spores in laboratory samples were enumerated using a membrane filtration method (Barbeau et al. 1997). Decimal dilution series of the laboratory samples were prepared using 10 mL of aliquot and 90 mL of sterile peptone dilution water (APHA AWWA WEF 1992). For each analysis, 2 to 4 dilutions were filtered through pre-sterilized 47 mm × 0.45 micron membrane filters (part no. 66586, Gelman Sciences, Ann Arbor, MI) using a vacuum filtration apparatus. The filters were then placed on sterile trypticase soy broth saturated pads in Petri dishes. The dishes were placed in a dry air oven at 70 to 75°C for 20 minutes, to inactivate any vegetative cells, and then into a 35°C incubator. After 22 to 24 hours, the plates were removed from the incubator and the number of colonies on each counted to determine the concentration of viable spores in the original sample in colony-forming units (CFU) per mL. For each laboratory sample, the entire enumeration procedure, including sample collection, preparation of the dilution series, and plating was done in triplicate. The logarithm of the geometric mean of the spore concentration in triplicate analyses was reported as log *N*.

***Cryptosporidium parvum* Methods**

Oocyst Production

Cryptosporidium parvum oocysts used in this study were originally obtained from Dr. Harley Moon (National Animal Disease Control center) and are known as the Iowa strain. Previously established methods of *Cryptosporidium parvum* oocyst production and purification from Holstein calves were used for the experiments (Finch et al. 1994; Finch et al. 1995; Finch et al. 1997). Calves, aged 2 to 4 days were infected with *Cryptosporidium parvum* oocysts and maintained on a diet of electrolyte solution. Feces collected from the calves at the onset of scouring were first passed through a series of sieves (400 to 75 μm). Oocysts were purified from the sieved feces by cesium chloride gradient centrifugation (Kilani and Sekla 1987). The calf diet of electrolyte solution reduced the lipid content of the feces and eliminated the need for the sucrose centrifugation pre-purification step that was used in previous work. Stock suspensions of purified oocysts were stored at 4°C in deionized water with antibiotics (100 μg/mL streptomycin, 100 μg/mL gentamicin, 100 U/mL penicillin) and 0.01% Tween 20 (Fisher Scientific).

Determination of Oocyst Infectivity in CD-1 Neonatal Mice

A mouse model using outbred neonatal CD-1 mice (Finch et al. 1993b) was used to evaluate infectivity of both fresh *Cryptosporidium parvum* oocycts and oocysts from ozonation experiments. Samples of oocysts suspended in water from ozonation experiments were centrifuged at 27,000 × *g* for 10 min at 4°C to concentrate the oocysts for inoculation into the neonatal CD-1 mice. The supernatant was aspirated and the cell pellet re-suspended in deionized water. Oocysts were counted in quadruplicate using a hemacytometer and appropriate dilutions prepared in de-ionized water for mouse infection. Breeding pairs of outbred CD-1 mice were obtained from the Charles River Breeding Laboratories (St. Constant, Quebec, Canada). The animals were given food and water *ad libitum* and were housed in cages with covers fitted with a 0.22 μm filter in a specific pathogen-free (P-2 level) animal facility. Mice were inoculated intragastrically 5 days after birth with a known number of oocysts suspended in 50 μL of deionized water. Intragastric inoculation was preformed using a ball-point neonate feeding needle (24 gauge, Popper and Sons Inc.) attached to a tuberculin syringe. For each laboratory sample, 2 to 4 dilutions of oocysts were prepared and cohorts of 5 to 10 mice were inoculated. Processing of the samples and inoculation of animals occurred within 24 h of laboratory exposure to ozone.

The infectivity of the oocysts was determined 7 days after infection. The mice were killed by cervical dislocation and the large intestine (rectum to 30 mm anterior to the caecum) was removed and placed in 10 mL of ultrapure water from a Milli-Q® system water. The intestine was homogenized for 45 to 60 s in a Sorvall Omni-Mixer and the homogenate collected into a 15 mL polypropylene test tube. The suspension was centrifuged at 2000 × *g* for 15 min. The supernatant was then removed and the pellet was re-suspended in 10 mL of Milli-Q® water containing 0.01% Tween 20, and was centrifuged at 2000 × *g* for 15 min. After the second centrifugation, the supernatant was discarded and 20 μL of the viscous pellet was removed and placed into a 6 mL polystyrene flow cytometer test tube fitted with a 35 μm sieve (Becton Dickinson, Franklin Lakes, NJ). The intestinal homogenate was forced through the sieve by adding 400 μL of 1% bovine serum albumen (BSA) in phosphate buffered saline (PBS) (Becton Dickinson). Samples were allowed to incubate for 15 min at room temperature to block non-specific absorption of the monoclonal antibody. One hundred microlitres of a 1:400 dilution of fluorescein labeled anti-*Cryptosporidium parvum* oocyst monoclonal antibody (ImmuCell) diluted in 1% BSA was subsequently added to each sample. The mixture was incubated at 37°C for 30 min.

The resulting suspension was examined for the presence of parasites on a FACScalibur flow cytometer (Becton Dickinson). Settings for the flow cytometer were as follows:

- forward side scatter – photodiode voltage equivalent to E00, AmpGain 4.00;
- side light scatter – photomultiplier voltage set to 402, AmpGain 4.00; and
- FL1 – photomultiplier voltage set to 470.

All flow cytometric analyses were done at a high flow rate using PBS as the sheath fluid. Fifty-thousand events (parasite counts) were collected for each intestinal homogenate sample. Mice were scored as infected with *Cryptosporidium parvum* when the number of events segregating into a defined fluorescence region was greater than 1.25%. At regular intervals, flow cytometric results were confirmed using conventional microscopy methods. Daily negative controls were run in which the infectivity was checked in a group of mice that were not exposed to oocysts. None of the negative control mice were positive for infection. A complete discussion of the flow cytometric procedures, including a discussion of the rational for scoring mice as infected and a comparison to conventional microscopy, was presented by Neumann and co-workers (2000a, 2000b).

Interpretation of* Cryptosporidium parvum *Infectivity Data

The proportion of mice positive for infection 7 days post-inoculation, *P*, was determined for each cohort using the previously described methods. The estimated number of infectious oocysts in the inoculum to each mouse, *d*, was then estimated using a logistic dose response model for *Cryptosporidium parvum* oocyst infectivity in the neonatal CD-1 mice. The form of the model is (Neter et al. 1989):

$$\ln\left[\frac{P}{1-P}\right] = \beta_0 + \beta_1 \log d \tag{2.5}$$

The parameters of the logit model, β_0 and β_1, were determined for the batch of oocysts used in the ozonation experiments in a series of oocyst dose-response experiments. Cohorts of 5 to 10 neonatal CD-1 mice were inoculated with levels of inoculums ranging from 25 to 200 oocysts per mouse. The number of animals positive for infection at each dose level was determined using the previously described methods. Parameters of the logit model were estimated from the results of the dose response experiments by maximizing the ln likelihood function, *Lh*, for binary data, given by (Brand et al. 1973)

$$\ln Lh = \sum_{i=1}^{\lambda} Y_i(\beta_0 + \beta_1 X_i) - \sum_{i=1}^{\lambda} \ln[1 + \exp(\beta_0 + \beta_1 X_i)] \tag{2.6}$$

The subscripts i = 1, 2, ... λ represent each individual mouse used in the dose response experiment, X_i is the log (base 10) of the inoculum size for each mouse, and Y_i is the binary score (0 = not infected, 1 = positive) of each mouse after 7 days. The likelihood function was maximized and the model parameters estimated using the Solver function in Microsoft Excel 2000 (Microsoft Co., Redmond, WA). Confidence intervals for the logit model parameters were calculated at the $1-\infty$ significance level by assuming that the errors of the model are normally distributed and determining the profile for maximum likelihood estimators given by (Seber and Wild 1989):

$$\ln Lh_{1\text{-}\alpha}(\beta) \geq \ln Lh_{\max}(\beta) - \frac{1}{2}\chi^2_{p,\alpha} \tag{2.7}$$

Here β_0 and β_1 are the logit response parameters, $\chi^2_{p,\alpha}$ is the chi-squared distribution, ln $Lh(\beta)$ is the likelihood function for normally distributed errors and ln $Lh_{max}(\beta)$ is the maximum of the likelihood function. Approximate 90% confidence intervals were calculated in Microsoft Excel 2000 by varying one parameter at time with the other set at the optimal estimate and satisfying the equality constraint of Equation 2.7.

Table 2.2
Results of dose response experiments with the batch of *Cryptosporidium parvum* oocysts used in the static mixer contactor experiments

Dose Response Experiment	Oocyst Age (days)	Inoculum Size *D*	No. of Mice in Cohort	No. of Mice Positive for Infection
1	16	25	10	3
		50	10	5
		100	10	9
		200	10	10
2	30	25	10	1
		50	10	5
		100	10	6
		200	10	10
3	44	25	10	6
		50	10	5
		100	10	6
		200	10	10
4	58	25	10	2
		50	10	4
		100	10	4
		200	10	8
5	72	25	10	4
		50	10	5
		100	10	5
		200	10	8

Characterization of Laboratory Oocysts

All oocysts used in this study were from a single batch and were used in experiments between 18 and 75 days after isolation. Five dose-response trials were conducted with these oocysts during this time period in order to determine the infectivity characteristics. The results of those individual experiments, summarized in Table 2.2, indicated that the oocysts maintained their infectivity during storage. The logit dose-response model was, therefore, regressed to the pooled dose-response data using the maximum-likelihood method and the logit model parameters generated represent the characteristics of the oocysts averaged over the 75 day period. The results are shown in Figure 2.5. Best fit logit model parameters and the corresponding 90% confidence intervals were:

$\beta_0 = -5.09\ (-5.3\ ,\ -4.8)$ and $\beta_1 = 2.97\ (2.8,\ 3.1)$

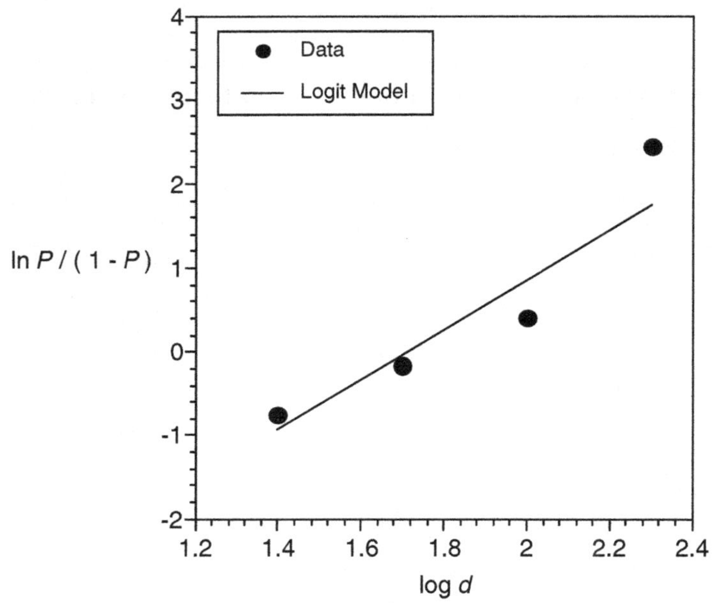

Figure 2.5 Dose-response of *Cryptosporidium parvum* oocysts used in laboratory trials and maximum likelihood fit of logit dose-response model

The estimated dose required for 50% infection in the mice (ID_{50}) can be calculated from these parameters as follows:

$$ID_{50} = \text{antilog}\ (-\beta_0 / \beta_1) = 52.$$

In comparison, the ID_{50} of oocyst batches used to generate a kinetic model for inactivation by ozone at 22°C in earlier work ranged from 60 to 347 (Gyürék et al. 1999). Therefore, the oocysts used in this study were slightly more infectious than those used in the earlier work.

Giardia muris Methods

Cyst Production

The strain of *Giardia muris* used in this study was originally isolated from a golden hamster by Roberts-Thomson et al. (1976). The parasite has been maintained in the laboratory since 1981 using bi-weekly passages in CD-1 or C3H/HeN mice. Methods for mouse inoculation and subsequent cyst collection and purification from mouse feces were adapted from those used previously (Belosevic and Faubert 1983; Finch et al. 1992; Labatiuk et al. 1991). Cysts for experiments were produced by inoculating groups of 8 to 11 week-old male C3H/HeN mice with purified suspensions of fresh cysts. Mice were inoculated by gastric intubation. Feces from the infected mice were collected over a 2 to 3-h period on each day of a three-day period. Collection periods were between 4 and 28 days post-infection. To isolate cysts, the daily fecal samples were emulsified in Milli-Q® water, layered on 1 M sucrose solution and then centrifuged for 15 minutes at 400 × *g* and 4°C. After centrifugation, the cyst rich layer at the water-sucrose interface was carefully collected and then washed 3 to 4 times in Milli-Q® water by centrifugation at 600 × *g* for 10 min. Cyst concentration was then determined by hemocytometer count. Cysts for experiments were stored in Milli-Q® water at 4°C and used within 48 h of collection.

Interpretation of Cyst Infectivity in Mice

The pattern of *Giardia muris* cyst shedding in mice is characterized by a latent period, in which no or few cysts are observed in the feces. The latent period is followed by a rapid increase in cyst output until a stable output of about 10^6 cysts per g of feces is reached, regardless of the size of infective inoculum (Belosevic and Faubert 1983; Belosevic et al. 1984). The feature that distinguishes the size of the infective inoculum is the duration of the latent period and the position in time of the rapid increase phase. To determine the relationship between inoculum size and latent period, cohorts of five, 8 to 11 week old male C3H/HeN mice were infected with 10^5, 10^4, 10^3, 10^2, or 10 cysts each in dose-response experiments. Cysts for the dose-response were collected and purified on the same day as infection. Mice were housed together, but were transferred to individual false bottom cages for 2 h each day post infection for feces collection. The wet feces sample collected from each mouse was weighed and cysts were recovered and purified using the previously described sucrose gradient centrifugation method. After centrifugation, the cysts were washed once before counting on a hemocytometer. The average cyst output per gram of wet feces was determined for each cohort on successive post-infection days until cyst output was stable at about 10^6 cysts per gram of wet feces.

The cyst shedding patterns determined for successive oral inocula in a typical dose-response experiment are plotted in Figure 2.6. The latent period (*LP*) in days, for each inoculum size was defined as the arithmetic average of the times to reach cyst outputs of 10^3, 10^4 and 10^5 cysts per g of wet feces. These times were interpolated directly from the dose-response plots. A relationship between latent period and size of infective inoculum was developed using information from three separate dose response experiments performed as part of this work, as well as information from similar experiments conducted by Labatiuk (1992) using identical cyst collection methods. The information from these independent dose-response experiments indicates that the dose-response relationship was reproducible and was independent of the batch of cysts (Figure 2.7). The method of least-squares was used to regress the following linear model to the dose-response data:

$$\log(d) = (-10.4 \pm 0.5)\log(LP) + (9.5 \pm 0.3) \qquad (2.8)$$

In this model, *d* was the number of infectious cysts in the mouse inoculum and *LP* was the latent period. In Equation 2.8, the model parameter values were shown in brackets along with the standard errors.

Cysts from ozonation experiments were concentrated by centrifugation (800 × *g* for 10 min at 4°C). The supernatant was removed to about 2 mL. The cyst concentration in the remaining liquid was determined by minimum of eight hemocytometer counts. For each sample, 5 mice were inoculated with 0.2 mL of the concentrated suspension. Cyst production in the feces was monitored as described for the dose-response experiments and the latent period, *LP*, was determined in the same way. The number of infectious cysts in the inoculum to each mouse, *d*, was estimated from Equation 2.8.

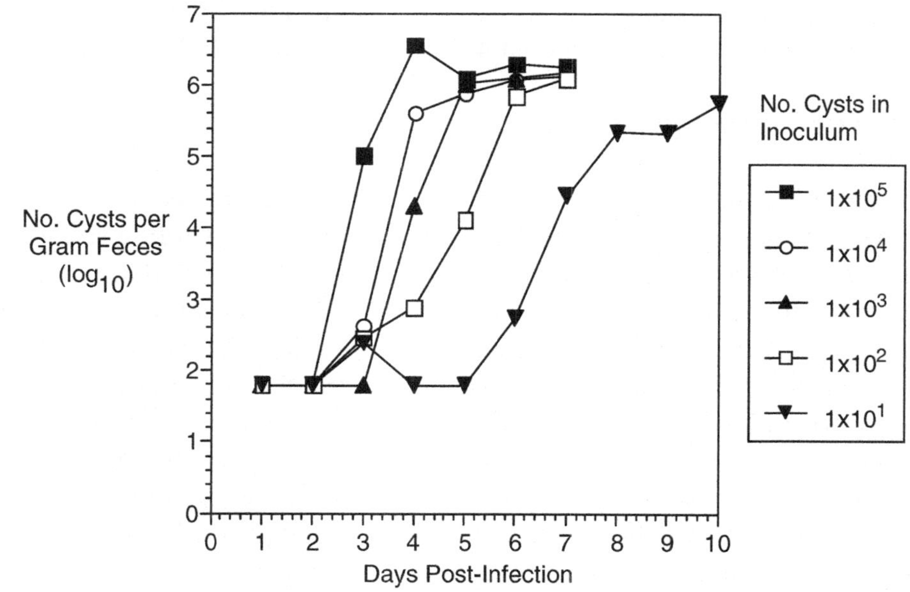

Figure 2.6 Pattern of cyst shedding in the faeces of C3H/HeN mice infected with fresh *Giardia muris* cysts in a dose-response experiment

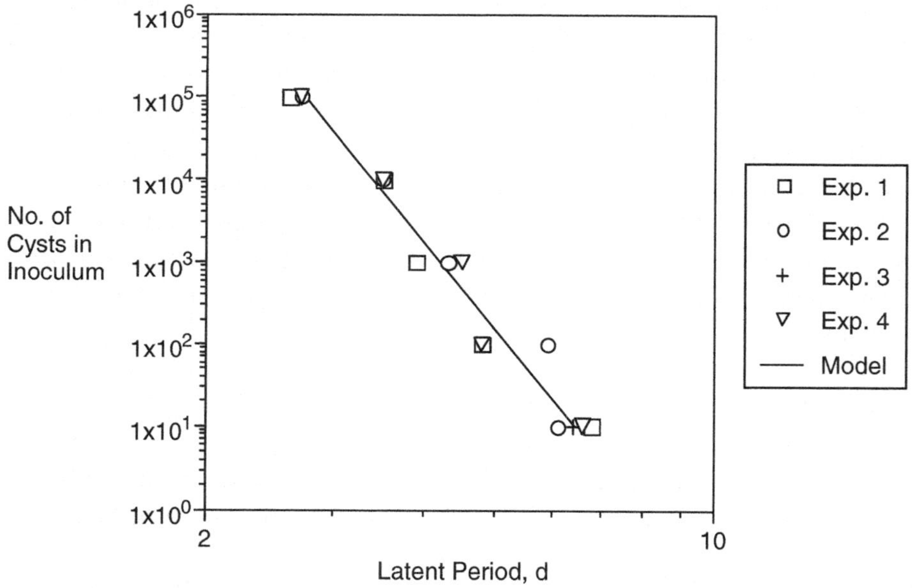

Note: Results from four dose response experiments are shown, including data from Labatiuk (1992) (indicated as Exp. 3).

Figure 2.7 Latent period (*LP*) in the C3H/HeN mice as a function of the number of fresh *Giardia muris* cysts in the oral inoculum (*d*)

CALCULATIONS

Mean Residence Time

The mean residence, t_m, at each sample point was estimated from the measured residence time distribution using the method of moments as described by Levenspiel (1972). The mean residence time, t_m was calculated from the exit age distribution function, $E(t)$, according to:

$$t_m = \int_0^\infty tE(t)dt \tag{2.9}$$

Equation 2.9 was evaluated numerically on a Microsoft Excel 2000 (Microsoft Co., Redmond, WA) spreadsheet program using the trapezoid rule with a step size of 1 s.

Average Concentration-Time Product

The microorganism inactivation measured for different levels of ozone treatment in the static mixer ozone contactor was interpreted using the product of the average ozone concentration and the effective contact time. The $C_{avg}t_m$ in the reactive flow segment was estimated from an integrated average of the dissolved ozone concentration profile. A first-order ozone decay rate in the reactive flow segment was estimated by fitting the following equation to the measured ozone concentration profile:

$$C = C_0 \exp(-k_d t_m) \tag{2.10}$$

where C was the predicted dissolved ozone concentration at each sample location, C_0 was the measured ozone concentration at the inlet of the reactive flow segment, t_m was the mean residence time and k_d was the first-order ozone decay coefficient. An example of a typical dissolved ozone profile in the reactive flow segment is shown in Figure 2.8 along with the fit of Equation 2.10. Non-linear least squares regression of Equation 2.10 to the measured profiles was done using the Solver function in Microsoft Excel 2000. An integrated average ozone concentration was then estimated for each sample point using the following equation (Langlais et al. 1991):

$$C_{avg} = \frac{C_0}{k_d t_m}\left[1 - exp(-k_d t_m)\right] \tag{2.11}$$

The contribution of the gas/liquid separator to the total $C_{avg}t_m$ was estimated using the following equation:

$$C_{avg}t_m = \left(C_{in} \times C_{out}\right)^{1/2} t_m \tag{2.12}$$

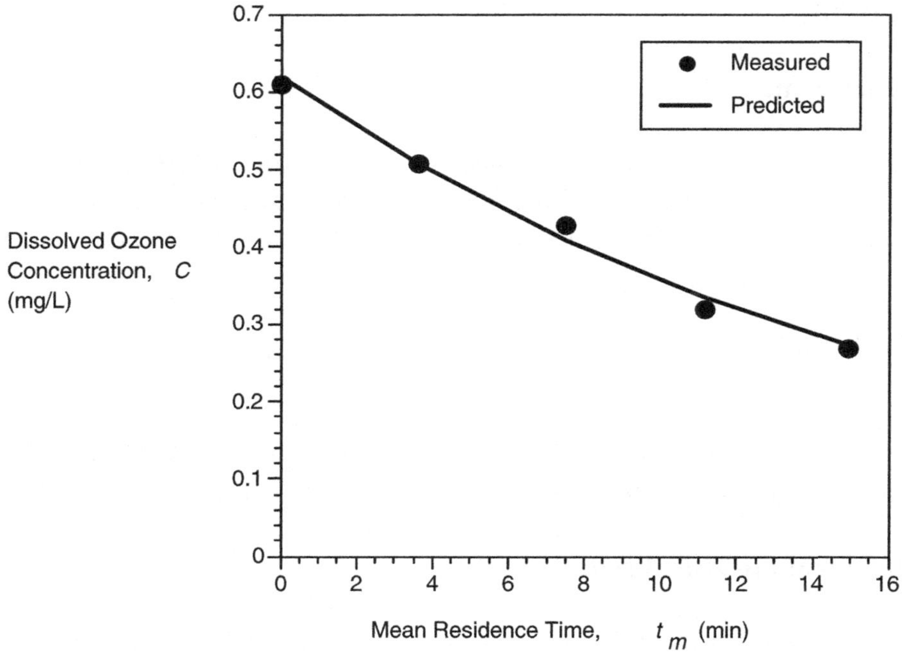

Note: Experiment was conducted at 16°C using *Bacillus subtilis* spores.

Figure 2.8 Typical ozone concentration profile in the reactive flow segment of the static mixer ozone contactor and fit of the first-order decay model

where C_{in} and C_{out} were the dissolved ozone concentrations at the inlet and outlet of separator, respectively, and t_m was mean residence time.

For batch reactor experiments, the $C_{avg}t_b$ product was estimated in a similar way except that batch time, t_b, replaced the mean residence time, t_m, in Equation 2.11. A typical concentration profile measured in a batch reactor experiment is shown in Figure 2.9.

Microorganism Inactivation

Microorganism inactivation in the static mixer contactor and in the batch reactors was interpreted in terms of the log inactivation ratio, *I*, which was defined by the following equation:

$$I = -\log \frac{N}{N_0} \tag{2.13}$$

where *N* and N_0 were the concentrations of live microorganisms in the water before and after the specified ozone treatment. The precise definition of *N* and N_0 varied according to the microorganisms studied. For laboratory trials with *Bacillus subtilis* spores, log *N* and log N_0 were determined directly from spore enumerations and were defined as follows:

log N_0 = base 10 logarithm concentration (CFU/mL) of the geometric mean concentration of live spores in the static mixer contactor feed stream prior to contact with ozone, or in the batch reactor control flask.

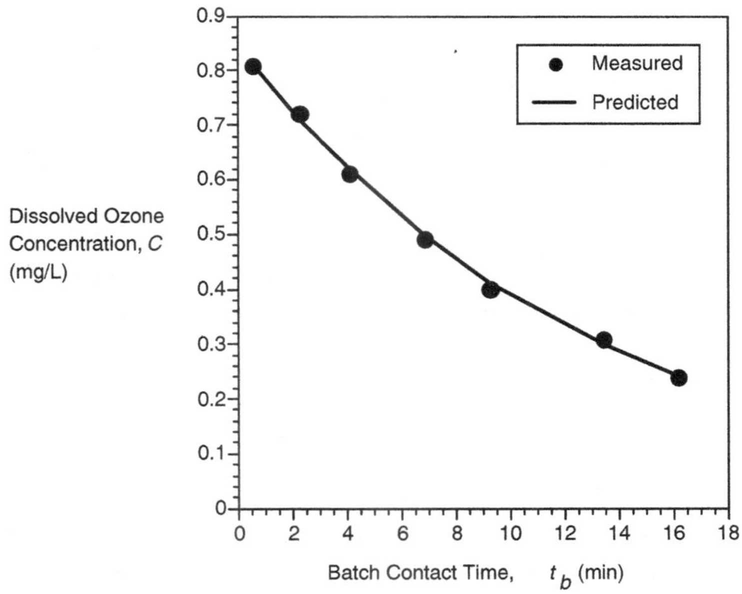

Note: Experiment was conducted at 16°C using *Bacillus subtilis* spores.

Figure 2.9 Typical ozone concentration profile in a batch reactor experiment and the fit of the first-order decay model

log N = base 10 logarithm of the geometric mean concentration of live spores (CFU/mL) after the specified contact with ozone in the static mixer contactor or in the batch reactor laboratory flask.

For triplicate plate counts (N_1, N_2 and N_3) an average value of log N was determined as follows:

$$(\log N)_{avg} = 1/3\ [(\log N_1 + \log N_2 + \log N_3) \tag{2.14}$$

The standard error of log N and log N_0 and the Student t-distribution were used to construct 95% confidence intervals on the log inactivation ratio. The standard error, SE, of – log N/N_0 was determined as follows:

$$\text{SE}\ [-\log N/N_0] = \text{SE}\ [\log N_0] + [\text{SE}\log N] \tag{2.15}$$

The log inactivation ratio of the encysted protozoans was inferred from the outcomes of the infectivity analysis on the recovered cyst or oocyst samples. The inactivation ratio was estimated from:

$$I = -\log\left(\frac{N}{N_0}\right) = -\log\left(\frac{d}{d_0}\right) \tag{2.16}$$

In Equation 2.16, d and d_0 were defined as:

d = the estimated number of infectious cysts or oocysts in the inoculum to each mouse

d_0 = the total number of cysts or oocysts in the same inoculum as determined by hemocytometer count.

Standard errors and confidence intervals for the log inactivation ratio of the protozoa were not estimated because reported inactivation ratios were based on a single infectivity measurement.

RESULTS AND DISCUSSION

Residence Time Distribution Analysis

In order to accurately determine $C_{avg}t_m$ products for the designed experiment, the mean residence time, t_m, at the various sampling locations in the laboratory contactor was first determined in pulse tracer tests with methlyene blue dye. Examples of the exit age distribution, $E(t)$, measured at five sample locations are shown in Figure 2.10. These tracer tests were conducted in either duplicate or triplicate and the mean residence time, t_m, was calculated using Equation 2.9. The results of replicate t_m determinations are summarized in Table 2.4. These values of t_m were then combined with the dissolved ozone concentration profile measured in each spore challenge trial and the $C_{avg}t_m$ products were calculated according to Equations 2.10, 2.11 and 2.12.

The calculation of $C_{avg}t_m$ assumed perfect plug flow type behavior in the contactor. The skewed shape of the $E(t)$ curves, with the steep slope on the right and the extended tail on the left, was more characteristic of streamline or laminar flow, rather than perfect plug flow. Visual observations during tracer testing and the relatively low Reynold's number within the reactive flow segment (approximately 800) were also consistent with laminar-type flow. Segregated flow analysis using the measured $E(t)$ and estimates of the first-order ozone decay rate, k_d, showed that the error in the calculated $C_{avg}t_m$ introduced by the assumption of plug flow was less than 5%.

Mass Transfer Efficiency in the Static Mixer Contactor

Laboratory Plan

A series of ozone transfer efficiency measurements were made to establish the design and operational factors that control ozone gas-liquid mass transfer in the laboratory study ozone static mixer contactor. Figure 2.11 depicts the components of the system under study, and shows the laboratory variables and the relevant sampling locations. The laboratory variables that determined the hydrodynamic conditions within the static mixer were the superficial liquid velocity, v_s and the gas-to-liquid volumetric flow rate ratio G/L. The transfer efficiency experiments were based on the measurement of ozone concentration in the off-gas leaving the top of the gas/liquid separator, Y_o, and, therefore, incorporated any contributions from the static mixer, the gas/liquid separator and any interconnecting tubing.

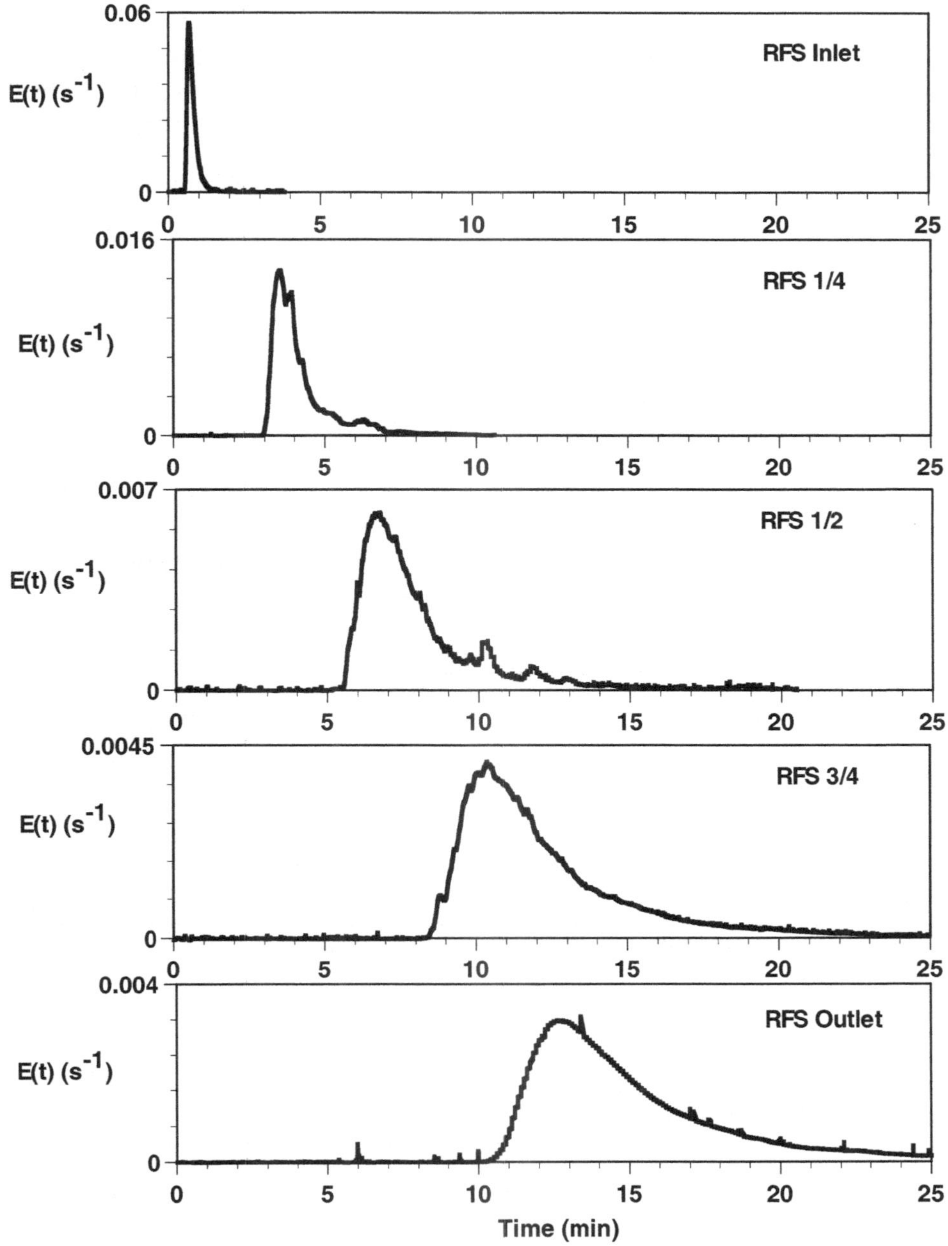

Figure 2.10 Exit age distribution, *E(t)* of the laboratory study static mixer contactor at five sampling locations along the length of the reactive flow segment

Preliminary Study

Preliminary experiments (8 trials) were conducted at a fixed feed water flow rate and with the gas/liquid separator adjusted to the maximum height of 2.3 m. At constant hydrodynamic conditions within the static mixer (i.e constant superficial velocity and *G/L*), transfer efficiency was essentially independent of the feed gas ozone concentration, Y_f, applied ozone dose, C_a, and the dissolved ozone concentration at the column outlet, C_0 (Table 2.5). Operation at lower *G/L* resulted in substantially greater transfer efficiency. Note, however, that *G/L*, ozone feed gas

Table 2.3
Target experimental conditions for the factorial designed experiment on the effect of static mixer hydrodynamic conditions on *Bacillus subtilis* spore inactivation

Experimental Setting	v_s m/s	G/L %	RFS Inlet Ozone Concentration, C_0 mg/L	RFS Flow Rate L/min
1	0.6	1.2	0.9 to 1.1	2.0
2	0.6	2.6	0.9 to 1.1	2.0
3	1.4	1.2	0.9 to 1.1	2.0
4	1.4	2.6	0.9 to 1.1	2.0

Table 2.4
Summarized results of replicate residence time determinations in the laboratory static mixer ozone contactor

	Mean Residence Time*, t_m (min)				t_{10}/t_m			
Sampling Location	1	2	3	Avg	1	2	3	Avg
RFS Inlet	0.63	0.72		0.68	0.68	0.69		0.69
RFS 1/4	4.5	4.3	4.5	4.4	0.83	0.78	0.83	0.81
RFS 1/2	8.0	8.6	8.6	8.4	0.75	0.73	0.75	0.74
RFS 3/4	12.2	11.9	12.1	12.1	0.77	0.79	0.78	0.78
RFS Outlet	16.0	15.5	16.5	16.0	0.80	0.77	0.82	0.80

* Mean residence times were determined for a feed water flow rate of 7.6 L/min.

Table 2.5
Results of preliminary trials with the laboratory-study static mixer system

Trial	v_s m/s	G/L %	Y_f g/NL	C_a mg/L	C_0 mg/L	TE %
1	0.93	4.3	47	2.0	1.6	80.2
2	0.93	4.4	86	3.7	3.0	80.9
3	0.93	4.3	10	0.4	0.4	77.9
4	0.93	4.3	71	2.4	2.8	78.3
5	0.93	1.3	90	0.9	1.0	92.2
6	1.04	1.2	104	1.1	1.6	92.7
7	1.04	1.2	20	0.2	0.2	93.8
8	1.04	1.2	104	1.1	0.9	93.2

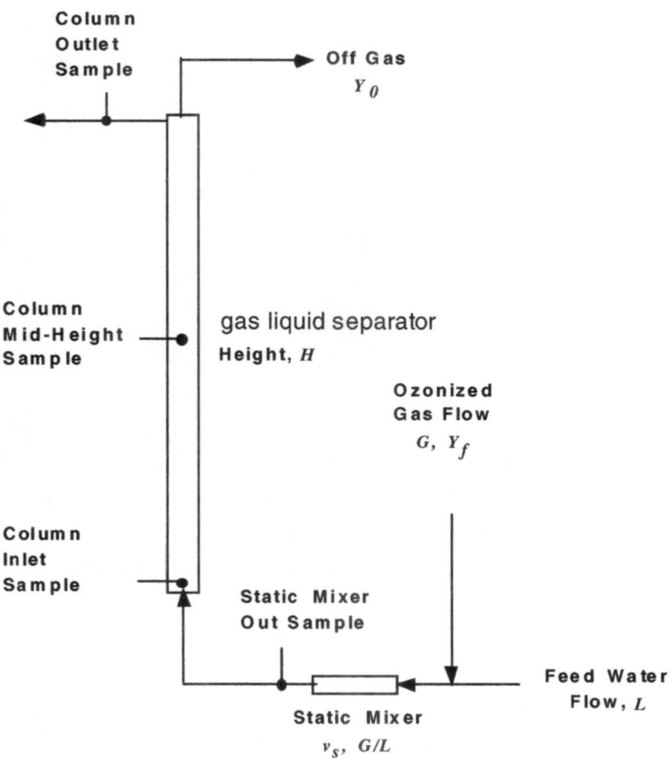

Figure 2.11 Illustration of ozone transfer efficiency measurements

concentration and dissolved ozone concentration were not independent variables in the static mixer system. The setting of one of these variables determined the other two. Therefore, it was difficult to determine to what degree the improved transfer efficiency at low G/L was caused by a change in the hydrodynamic conditions or by an increase in feed gas concentration. Either way, the results suggest that for a given ozone concentration, the most efficient operation is at a low G/L and high feed gas ozone concentration. This feature makes static mixer systems well-suited to the high concentrations and low gas flow associated with medium frequency ozone generators.

Transfer Efficiency Designed Experiment

The data suggested that transfer efficiency increases with increasing water velocity, decreasing gas liquid flowrate ratio and increasing height of liquid in the gas-liquid separator. To further study the relationship between the factors and the response variables, and to test any non-linear responses and interactions, a factorial experiment was designed to quantitatively evaluate the impact of hydrodynamic conditions within the laboratory static mixer constactor (v_s and G/L) on transfer efficiency. One of the advantages of a static mixer is that it provides rapid dissolution of the ozone and, therefore, reduces the requirement for large 2-phase contacting chambers. Examples of the measured dissolved ozone profile (Figure 2.12) in the static mixer and gas/liquid separator indicate that much of the ozone transfer in the laboratory contactor was completed by the time the 2-phase mixture entered the gas/liquid separator. In fact, a considerable component of the transfer occurred in the tubing that connected the outlet of the static mixer to the inlet of the gas/liquid separator. To better understand the contribution of the gas/liquid separator to ozone

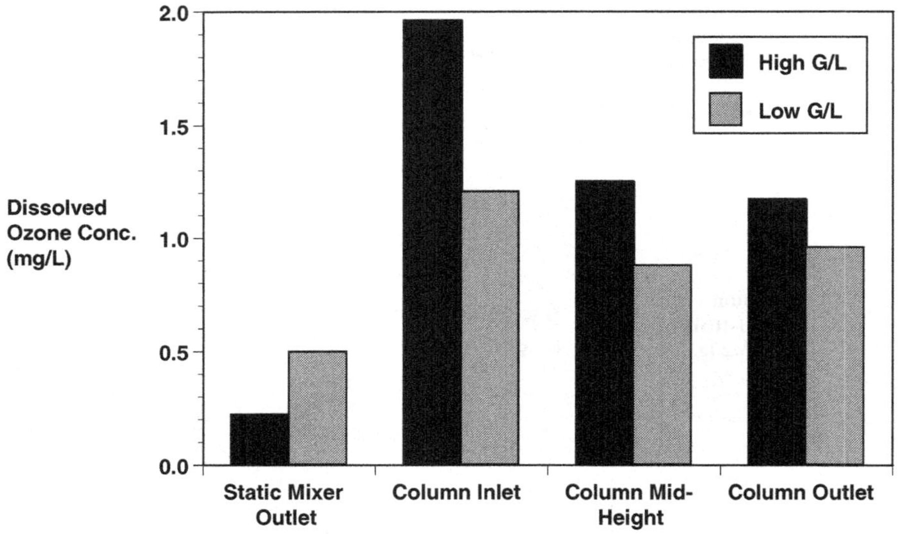

Figure 2.12 Typical dissolved ozone profiles in the laboratory study ozone static mixer contactor gas/liquid separator

Table 2.6
Transfer efficiency factorial designed experiment settings

Variable	Low Level	Medium Level	High Level
Coded Unit	-1	0	+1
v_S, m/s	0.7	0	1.3
G/L, %	1.2	2.7	4.1
H, m	0.46	1.36	2.26

gas-liquid transfer in the laboratory contactor, the height of the gas/liquid separator, *H*, was included as a third factor in the experimental design. The target levels for each of the three factors in the design are provided in Table 2.6. The response variable was the measured transfer efficiency. For each experimental setting of v_s and *G/L*, the objective was to maintain the ozone dose at a target of 1.2 mg/L by appropriate adjustment of the feed gas ozone concentration. Experimental trials were randomized to block out effects of day-to-day water quality variations.

The results of the experimental design trials were summarized in Table 2.7. Levels of the factors were shown in coded units (Box et al. 1978) and were adjusted for the actual conditions achieved during the laboratory trial. The coded unit of a given variable was defined as:

$$\text{Coded Value} = 2\left[\frac{\text{Measured Value - Low Level (-1)}}{\text{High Level(+1) - Low Level(-1)}}\right] - 1 \tag{2.17}$$

where the high and low levels of the variables are defined as in Table 2.6. Although the transfer efficiency was the primary response variable, the value of the ozone concentration at the inlet of the reactive flow segment, C_0, was also provided for information purposes. This was the same as the concentration of ozone at the outlet of the gas/liquid separator. During the course of the

Table 2.7
Results of ozone transfer efficiency factorial designed experiment on laboratory study of static mixer contactor

	Factor Values in Coded Units			Response Variables	
Trial	v_s	G/L	H	TE %	C_0 mg/L
16	-1.04	-1.02	-1	76.1	1.2
20	-1.15	-0.98	1	88.7	1.2
11	-1.21	1.22	-1	66.3	0.8
29	-0.94	0.85	1	77.4	0.8
17	0.04	-1.01	-1	86.7	1.0
31	-0.36	-0.88	-0.16	87.1	1.2
22	0.04	-1.01	1	93.0	1.1
15	0.04	0.95	-1	73.7	0.5
21	0.04	0.99	1	82.0	0.8
25	1.28	-1.02	-1	90.3	1.0
26	1.13	-0.99	-0.16	91.8	1.2
24	1.18	-1.01	1	93.8	1.3
10	1.10	-0.03	-1	81.3	1.0
23	1.18	-0.02	1	88.1	1.0
12	-0.16	0.05	-0.16	80.6	1.1
19	-0.09	0.00	-0.16	82.0	0.8
18	-0.08	-0.01	-0.16	80.7	0.9

experiment, it became clear that it was not possible to maintain stable operation of the system at the target test condition of high v_s and high G/L. A target condition of high v_s and intermediate G/L was substituted, resulting in a less than perfectly balanced factorial design.

To establish a mathematical relationship between transfer efficiency and the laboratory variables, a linear empirical model of the following form was fitted to the laboratory data of Table 2.7:

$$TE = \beta_0 + \beta_1 x_1 + \beta_2 x_2 + \beta_3 x_3 + \beta_{11} x_1^2 + \beta_{22} x_2^2 + \beta_{33} x_3^2 + \beta_{12} x_1 x_2 + \beta_{13} x_1 x_3 + \beta_{23} x_2 x_3 + \beta_{123} x_1 x_2 x_3 \quad (2.18)$$

Here x_1, x_2 and x_3 represented the levels of the laboratory variables v_s, G/L and H, expressed in coded units as per Equation 2.17. Coefficients of this model were calculated by multiple linear regression and using least squares minimization criteria. Computations were done on a Microsoft Excel 2000 spreadsheet program using the Solver function. Analysis of variance was used to reject terms that were not statistically significant at the 95% confidence level (i.e for which the probability of a type I error, p, was > 0.05). Several models with different combinations of

Table 2.8
Transfer efficiency model regression error analysis

	Degrees of freedom	Sum of Squares	Mean Square	F-statistic	p-value
Model Factors	5	878	176	126	2.5×10^{-9}
Error	11	15.3	1.4		
Total	16	893			

Model Term	Coefficient	Standard Error	p-value
Y-intercept	83.2	0.43	8.3×10^{-21}
v_s	3.9	0.37	3.9×10^{-7}
G/L	-6.1	0.38	4.7×10^{-9}
H	4.2	0.35	$1.2\ 10^{-7}$
v_s^2	-1.8	0.45	0.002
$v_s \times G/L \times H$	1.8	0.47	0.003

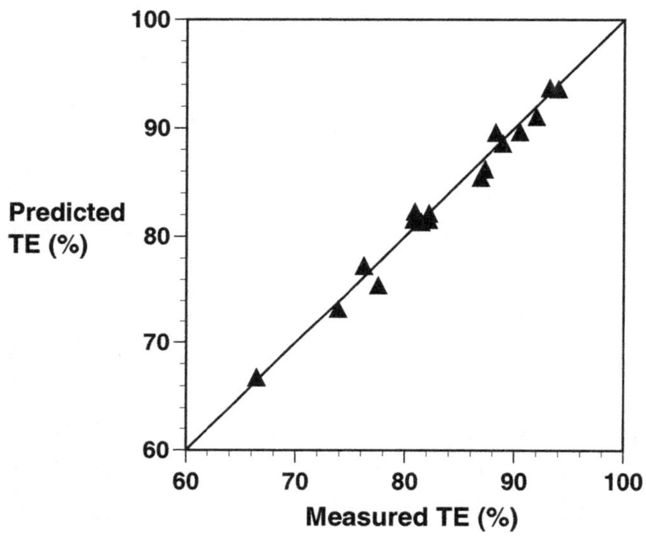

Figure 2.13 Transfer efficiency model fit

significant terms were tested. Of the various linear model forms explored, the following model was found to provide the best fit to the data:

$$TE = 83.2 + 3.9v_s' - 6.1(G/L)' + 4.1H' - 1.8v_s'^2 + 1.8v_s'(G/L)'H' \quad (2.19)$$

The prime (′) in Equation 2.19 indicated the value of the variable in coded units as defined by Equation 2.17. The F statistic and the p-values determined from the ANOVA showed that all terms in this model, as well as the overall model, were significant at the 95% confidence level (Table 2.8). The model was found to provide a good fit to the measured transfer efficiency (Figure 2.13).

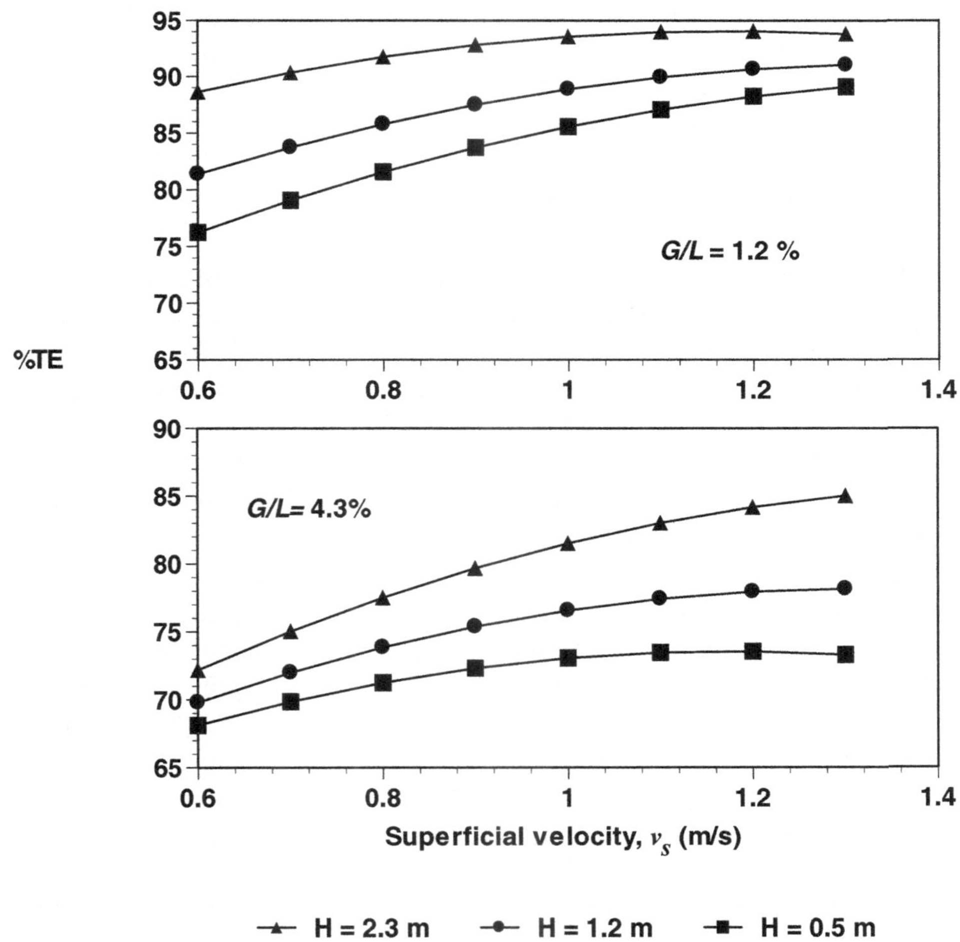

Figure 2.14 Examples of transfer efficiency model predictions: Effect of v_s, *G/L* and *H* on transfer efficiency at two levels of *G/L*

The interdependence of the system variables, indicated by the significance of the 3-factor interaction, was further explored by examining selected model predictions (Figure 2.14). Transfer efficiency was improved by increasing column height under all hydrodynamic conditions within the static mixer. The magnitude of the improvement, however, depended on static mixer hydrodynamics. At high v_s and low *G/L*, the hydrodynamics and ozone dissolution conditions were optimal and most of the transfer occured within the mixer itself. At this condition, the effect of the gas/liquid separator was relatively small. A five-fold increase in the height of the separator from 0.5 m to 2.3 m produced only a 4 percent increase in transfer efficiency. At a low v_s, but still at a low *G/L*, where the hydrodynamics of the static mixer are less conducive to efficient gas-liquid mixing and mass transfer, a much larger (~11 %) percent increase in transfer efficiency can be realized by making the column taller. Interestingly, at a high *G/L* condition the effect was reversed. The column height had less impact at low velocities but greater impact at high velocities.

The model was used to provide a rough estimate of the transfer efficiency of the static mixer itself by extrapolating to a condition of zero separator height. At a *G/L* of 1.2 % and velocity of 1 m/s the predicted efficiency of the overall system was 93.8 %. If the length of the separator is set to zero, i.e. the static mixer only, the model predicted a transfer efficiency of 88 %. Therefore, 94 % of the ozone transfer occured in the static mixer itself.

Inactivation of Microorganisms by Ozone

Disinfection Scale-Up Experiments

As the lab- and pilot-scale units were built at the University of Alberta, a set of "side-by-side" trials were completed there to determine the disinfection efficiency of both the static mixer units and to test scale-up of the disinfection process. *Bacillus subtilis* spores were used as the microbial indicator of disinfection efficiency. The two contactors were operated side-by-side in which the de-gassed effluent from the pilot "front end" (static mixer + gas/liquid separator) was split. The larger portion was fed to the pilot-scale contactors and the smaller portion was fed to the lab-scale contactors. The ozone inactivation characteristics of the spores were determined in a separate set of batch disinfection trials using the same test water. A total of 15 batch disinfection trials were completed at various combinations of ozone dose and contact time.

Spore inactivation as a function of the average ozone Ct product is shown in Figure 2.15 for each of the three contactors; the batch well-mixed contactor, the lab-scale plug flow contactor and pilot-scale plug flow contactor. The general character of the spore inactivation curve versus Ct was reproduced in each contactor. For these *Bacillus* spores the ozone inactivation curve included a distinct initial lag period or shoulder, followed by a curvi-linear increase in inactivation with Ct and then a tailing region at high Ct. With the exception of the tail area at high Ct, the inactivations in the three contactors are similar for a given contact time. There was no significant spore inactivation within the short contact time (< 0.2 s) of the static mixer itself, nor within the contact time of the gas/liquid separator (60 seconds). Measurable spore inactivation was observed only after several minutes of contact time.

Agreement between spore inactivation in the three types of contactors suggests that the disinfection process is independent of contactor size. The results showed that a level of disinfection equivalent to an ideal well-mixed batch system can be achieved in an ozone static mixer–plug flow type contactor system given the same average ozone concentration and contact time. Inactivation kinetics determined in well-mixed batch reactors can be used to predict performance in larger flow systems with plug-flow type hydraulics. In addition to demonstrating that disinfection results from batch well-mixed reactors can be transferred to larger plug flow systems, this study demonstrates one of the benefits of using the static mixer in an integrated disinfection process. The static mixer achieves rapid and efficient dissolution of ozone and achieves a uniform ozone residual in the water. Utilizing a static mixer, followed by off-gas separation and contacting with ozone in a plug flow type vessel, it is possible to achieve a level of microbial inactivation equivalent to an idealized batch well-mixed contactor.

These results showed that microbial inactivation between the pilot-scale and lab-scale contactors can be expected to be similar given the same Ct. This was important because protozoan disinfection trials were conducted only on the lab-scale contactor at the University of Alberta. The pilot-scale contactor performance were determined with various water qualities at several water treatment plants using *Bacillus subtilis*. The side-by-side test results formed the crucial scale-up link between the lab-scale and pilot-scale contactors. The results also confirmed that an integrated ozone disinfection process consisting of a static mixer, gas-liquid separator and a plug flow contactor can achieve microbial inactivation that is comparable to an idealized well-mixed batch

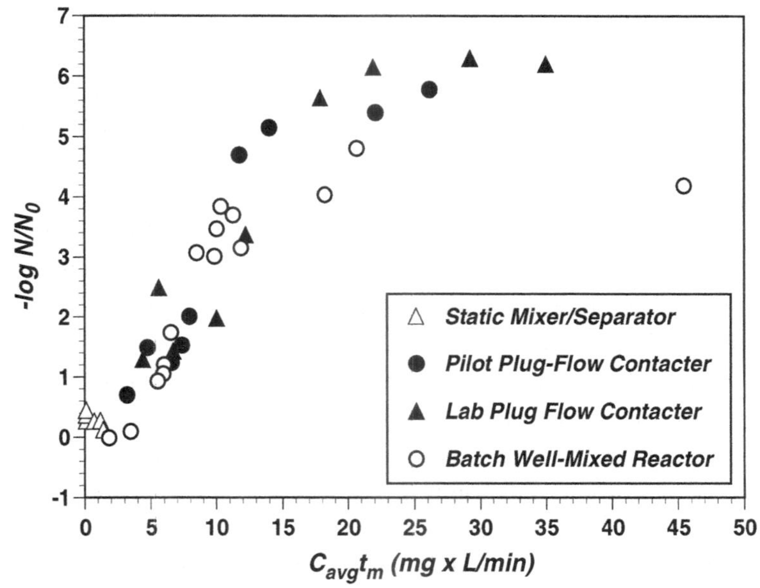

Notes: Batch experiments were done at 16 ± 1° C. Pilot- and lab-scale static mixer experiments were done at 13.9 °C to 18.2 °C, pH = 7.7.

Figure 2.15 ***Bacillus subtilis*** **spore inactivation as a function of average ozone Ct in lab-scale, pilot-scale plug flow contactors and in a batch well-mixed reactor, pH = 7.7**

reactor. No significant microbial inactivation was observed at the outlet of the static mixer or the gas-liquid separator.

Inactivation of *Bacillus subtilis* Spores in the Static Mixer Contactor

Experimental Plan

The ozone transfer efficiency experiments demonstrated that the superficial liquid velocity, v_s, and the gas-liquid flow rate ratio, *G/L*, within the static mixer had a significant and predictable effect on transfer efficiency. The objective of the experiments described in this section were to determine if these same hydrodynamic parameters had a measurable effect on microorganism inactivation and to determine if inactivation was enhanced by virtue of the hydrodynamic and gas-liquid contact conditions within the static mixer. Cultured *Bacillus subtilis* (ATCC 6633) spores were used as indicators of microorganism reduction efficiency in the challenge tests. A factorial experiment was designed using the laboratory study static mixer contactor in which the hydrodynamic conditions within the static mixer were varied by adjusting the settings of v_s and *G/L*. Target conditions for the factorial design are provided in Table 2.3. The conditions of dissolved ozone contact in the reactive flow segment (RFS) were maintained approximately constant at the various experimental settings of v_s and *G/L*. The response measured in each experimental trial was the level of spore inactivation as a function of the dissolved ozone $C_{avg}t_m$ product. All spore challenge trials were done using the shortest gas/liquid separator configuration (H = 0.45 m) and at a target water temperature of 16°C. Spore stock preparations B and C were used in all challenged experiments.

Results

Duplicate spore challenge experiments were done at each of the four static mixer hydrodynamic conditions listed in Table 2.3. A summary of the operating conditions actually achieved in each challenge experiment is provided in Table 2.9. The ozone transfer efficiency values listed in Table 2.9 were calculated using the previously derived empirical model (Equation 2.19) with the height of the gas/liquid separator, *H*, set equal to 0.45 m and v_s and *G/L* set at the values indicated in Table 2.9. The last column in Table 2.9 shows the spore inactivation within the static mixer itself, which was based on measurement of viable spore concentrations in the feed and at the outlet of the static mixer.

Spore inactivation was plotted as a function of the dissolved ozone $C_{avg}t_m$ product in the contactor for each of the four duplicate sets of challenge experiments in Figure 2.16. The $C_{avg}t_m$ values in this figure were based on dissolved ozone analysis by the indigo trisulphonate colorimetric method. During the experimental trials with *Bacillus subtilis* spores, the ratio of the ozone concentration measured by the indigo trisulphonate method to that measured by the UV absorbance method varied between 1.0 - 1.7. This large variability in the offset between the two analytical measurements was believed to have been due to day-to-day variations in the quality of the building tap water and fluctuation in the concentrations of compounds that interfered with the UV absorbance measurement. The baseline correction for the UV absorbance method described by Equation 2.4 was not used during the experiments with *Bacillus subtilis* spores and was only adopted in later experiments with *Cryptosporidium parvum* and *Giardia muris*. The indigo trisulphonate method was less sensitive to the presence of interfering compounds.

The results of all eight spore challenge experiments at the different static mixer hydrodynamic conditions are plotted together in Figure 2.17. The sigmoidal characteristic of the

Table 2.9
Laboratory conditions for *Bacillus subtilis* spore inactivation experiments in the laboratory-scale ozone static mixer contactor

Trial	v_s m/s	*G/L* %	C_0* mg/L	k_d† min^{-1}	*TE*‡ %	Temp. °C	pH	Inactivation in the static mixer§ log-units
34	0.63	1.3	1.22	0.039	77	16.6	7.6	0.30 ± 0.5
38	0.59	1.3	1.22	0.018	75	15.6	7.8	0.07 ± 0.2
35	1.34	2.5	0.88	0.016	82	16.0	7.6	0.01 ± 0.08
39	1.38	2.4	0.98	0.037	82	15.4	7.9	-0.03 ± 0.2
36	1.37	1.2	1.34	0.030	89	16.3	7.5	0.01 ± 0.5
40	1.40	1.2	0.83	0.032	89	16.4	7.8	0.11 ± 0.4
37	0.55	3.2	1.16	0.023	70	16.4	7.7	0.04 ± 0.2
41	0.69	2.5	1.47	0.039	75	16.4	7.8	0.03 ± 0.4

* Ozone concentration measured at the inlet of RFS.
† Estimated 1st order ozone decomposition constant in RFS.
‡ Ozone transfer efficiency estimated from empirical model.
§ Mean and ± 95% confidence interval shown.

ozone inactivation curve that was observed with this strain of spores in the batch reactor trials was also apparent in the inactivation curves generated in the static mixer. There was no discernable difference between the inactivation curves determined at the various combinations of v_s and G/L. The box plot of Figure 2.18 describes the distribution of the measured inactivation within the static mixer itself that was measured in the eight trials. The average inactivation measured within the static mixer itself was 0.07 log-units. A two-tailed Student t-test indicated that this difference was not statistically significant at the 95% level (probability of a type I error, $p = 0.05$). The close agreement between inactivation curves and the negligible inactivation across the static mixer was evidence that spore viability was neither directly nor indirectly affected by the hydrodynamic and gas-liquid contacting conditions within the static mixer. Rather, spore inactivation was determined exclusively by the degree of dissolved ozone contact.

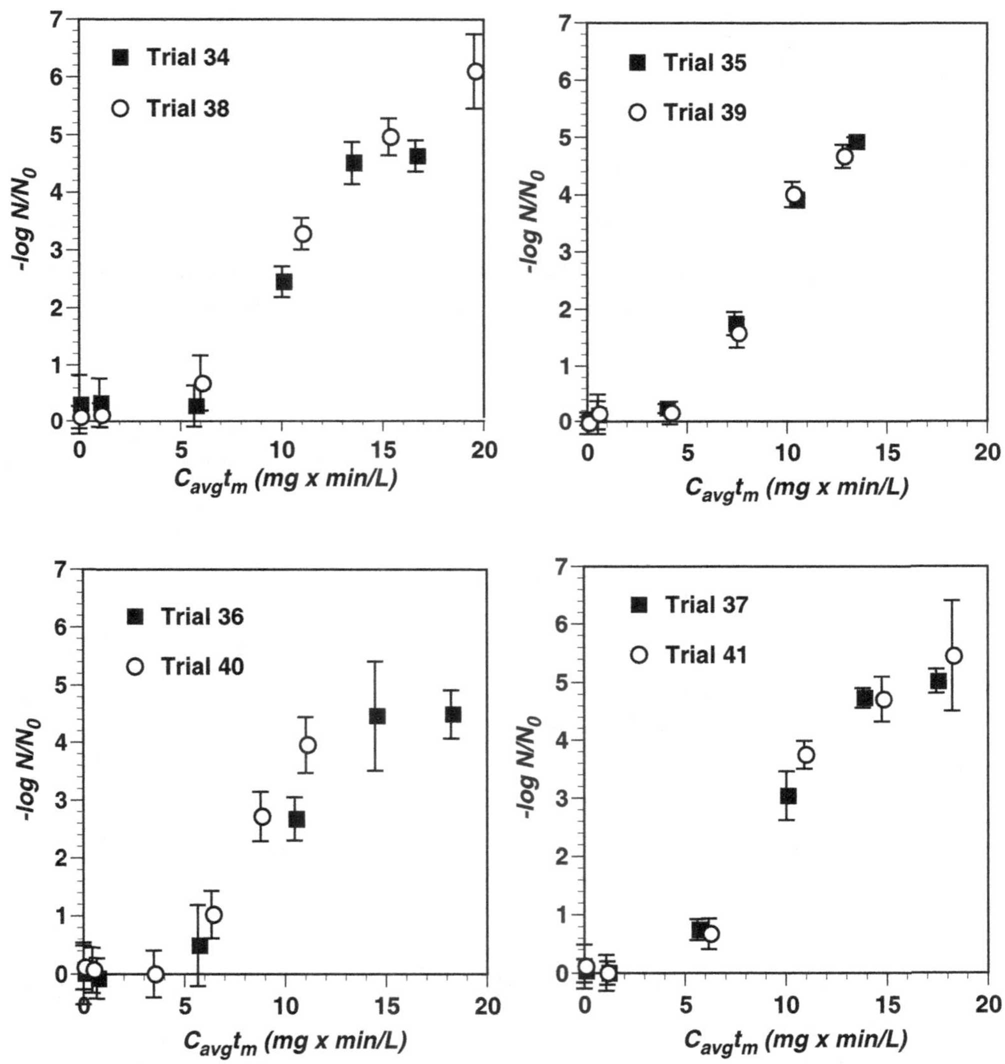

Note: Spore inactivation is shown as function of the dissolved ozone $C_{avg}t_m$ product at various sampling locations in the static mixer contactor. Ozone concentration is based on the indigo trisulphonate measurement method.

Figure 2.16 Results of duplicate challenge trials with *Bacillus subtilis* spores at different static mixer hydrodynamic conditions

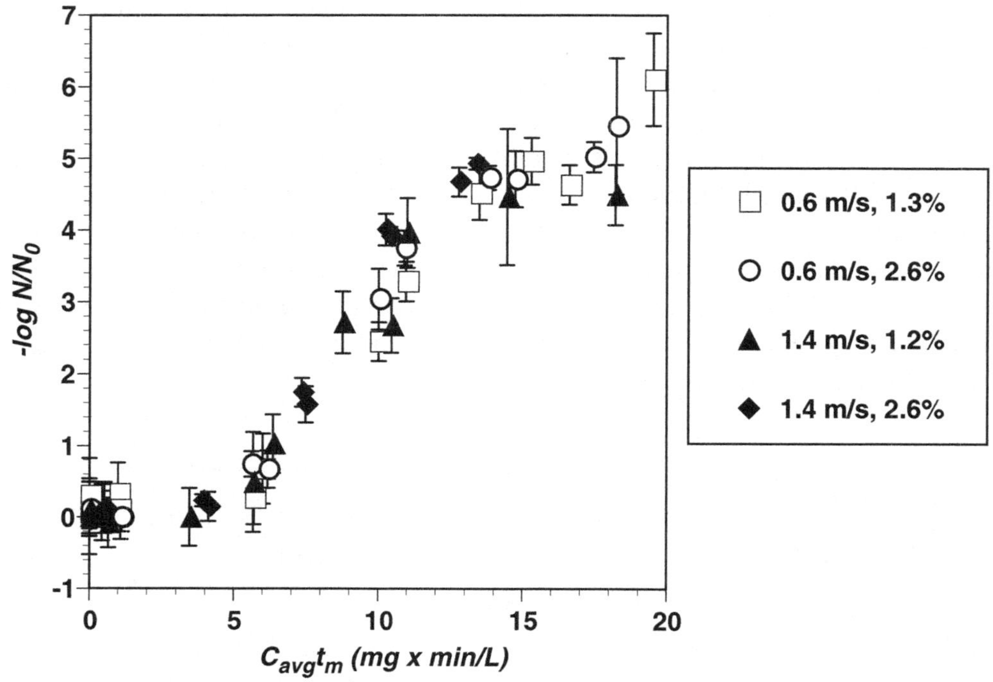

Figure 2.17 Comparison of *Bacillus subtilis* spore inactivation measured in the static mixer contactor at various static mixer different hydrodynamic conditions (v_s and *G/L*)

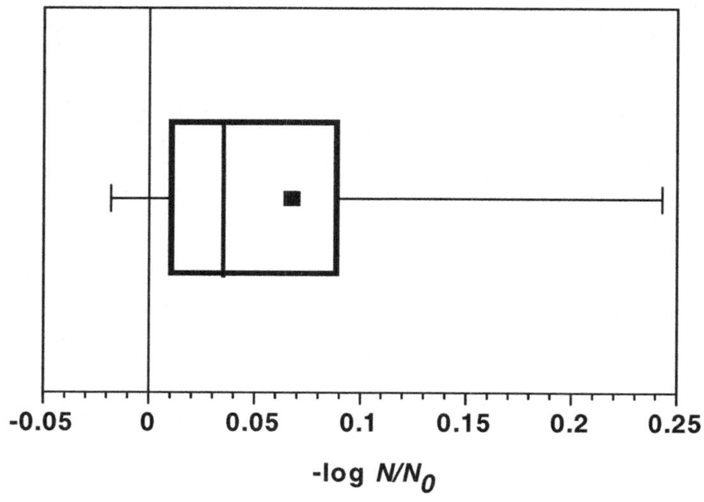

Note: From left to right, lines represent 10%, 25%, median, 75% and 90% percentile boundaries, respectively. The square symbol represents the mean.

Figure 2.18 Box plot of measured *Bacillus subtilis* spore inactivation across the static mixer itself

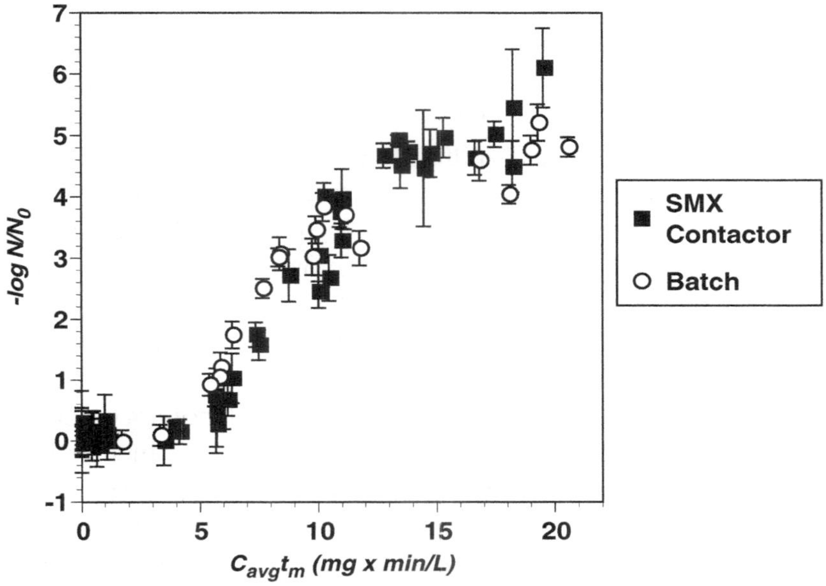

Figure 2.19 Comparison between *Bacillus subtilis* spore inactivation measured in the static mixer contactor at all hydrodynamic conditions (v_s =0.7 to 1.33 m/s, G/L=1.2 to 2.6%) and spore inactivation measured in batch reactors

The inactivation curves measured at the various hydrodynamic conditions within the static mixer contactor were found to agree closely with the inactivation curve determined in a series of batch reactor experiments (Figure 2.19).

Care was taken that the conditions of these batch reactor experiments closely matched those of the challenge experiments with the static mixer contactor, with the exception the method of ozone contact. Trials were done at 16°C and the indigo trisulphonate method of ozone measurement was used to interpret the results. Spore samples from the same preparations that were used in the static mixer contactor experiments (preparations B and C) were used in these batch reactor experiments. During the experimental trials, samples of water were collected from the feed tank of the static mixer contactor. These were combined, stored at 4°C for no longer than 4 weeks, and then used as the water matrix in the batch reactor trials. The only apparent effect of contact within the static mixer was a slight reduction in the degree of tailing in the spore inactivation curve (Figure 2.19). It was hypothesized that the high liquid shear condition within the static mixer (micromixing) may have promoted further break-up the laboratory spore suspensions compared to the vortex mixing procedure used in the batch reactor trials, leading to the observed reduction in tailing.

Inactivation of *Cryptosporidium parvum* in the Static Mixer Contactor

Laboratory Plan

Nine challenge experiments with *Cryptosporidium parvum* oocysts were completed with the laboratory study static mixer ozone contactor. For each, the short gas/liquid separator configuration (H = 0.45 m) was employed and the feed water temperature was regulated to 22°C.

The length of the reactive flow segment was varied between laboratory trials to provide total contact times of approximately 4, 8 and 16 min. Samples of feed water (after addition of oocysts but before ozone addition) and of the reactive flow segment outlet water were collected for oocyst recovery and infectivity analysis with the CD-1 neonatal mouse assay. Seven trials were done at a superficial liquid velocity, v_S, of 0.6 m/s in the static mixer. Two other trials were done at a v_S of 1.4 m/s to investigate the potential effect of static mixer hydrodynamics on oocyst inactivation. The *G/L* was maintained between 1.2% and 1.6%. In five of the seven trials at $v_S = 0.7$ m/s, an additional sample was collected immediately after the ozone static mixer to determine the extent of oocyst inactivation within the static mixer itself. A single control trial was done using oxygen gas in place of ozonized gas and with a total contact time of 16 minutes.

In addition to the static mixer contactor challenges, a set of seven batch reactor trials were also carried out to determine if the ozone resistance of the laboratory oocysts in the challenge experiments had a significant influence on the ozone inactivation kinetics. For these batch trials, sub-samples of oocysts from the same preparations used in the static mixer contactor experiments were suspended in samples of feed water from the static mixer apparatus. The suspensions were then exposed to various dissolved ozone conditions in batch reactors at a controlled temperature of 22°C. Details of the batch reactor trials are provided in Table A.1 of Appendix A.

Contactor Modeling

To assess microorganism inactivation efficiency in the static mixer contactor, the measured oocyst inactivation was compared to the inactivation predicted by mathematical models. Two approaches were used to model inactivation in the static mixer contactor; plug flow analysis (PFA) and segregated flow analysis (SFA). Both approaches assume that the rate of oocyst inactivation is determined by contact with dissolved ozone, and can be described by kinetic models previously developed in batch reactor experiments. The two approaches differed only in how the macro-mixing pattern and the residence time distribution of the contactor is considered. In plug flow analysis, the contactor was assumed to behave as an ideal plug flow reactor. Segregated flow analysis is a chemical reaction engineering technique in which batch reaction kinetics are integrated with residence time distribution information to estimate conversion in a non-ideal flow reactor (Levenspiel 1972). Segregated flow analysis has been used by others to model microorganism reduction by chlorine in water and wastewater treatment (Haas 1988; Lawler and Singer 1993; Trussell and Chao 1977) and is an integral component of the integrated disinfection design framework (IDDF) approach to analyze inactivation of microorganisms in drinking water treatment (Bellamy et al. 2000). The general equation for segregated flow analysis is:

$$\frac{N}{N_0} = \int_{t=0}^{t=\infty} \left(\frac{N}{N_0}\right)_{batch} E(t)dt \tag{2.20}$$

where N/N_0 is the is the estimated inactivation ratio of the contactor, $(N/N_0)_{\text{batch}}$ is the inactivation ratio estimated from batch kinetics, and $E(t)$ is the exit age distribution typically determined in tracer tests.

The kinetics of *Cryptosporidium parvum* oocyst inactivation by dissolved ozone were interpreted using the Hom-type kinetic model developed by Gyürék and others (1999). The basic form of the Hom kinetic model is:

$$\frac{dN}{dt} = -kmNC^n t^{m-1} \tag{2.21}$$

In logarithm base 10 integrated form, with constant ozone concentration, C, Equation 2.21 becomes:

$$-\log\frac{N}{N_0} = k'C^n t^m \tag{2.22}$$

where $k' = k/\ln(10)$.

For the case of first-order ozone decomposition, Equation 2.21 can be integrated and solved exactly using the incomplete gamma function to yield the incomplete gamma Hom formulation (Gyürék and Finch 1998; Haas and Joffe 1994):

$$-\log\frac{N}{N_0} = \frac{mk'C_0}{(nk_d)^m}\gamma(m, nk_d t) \quad m > 0,\ nk_d t \geq 0 \tag{2.23}$$

where $\gamma(m, nk_d t)$ is the incomplete gamma function.

Using batch reactors with Iowa strain oocysts suspended in ozone demand-free phosphate buffered lab water at 22°C, Gyürék and co-workers (1999) determined the following best-fit values for the model kinetic parameters: $k' = 0.68\ \text{min}^{-1}$, $n = 0.70$ and $m = 0.73$. Dissolved ozone was measured using direct UV absorbance at 260 nm with a molar extinction coefficient of 3,300 $M^{-1}cm^{-1}$, and infectivity reduction was measured using a neonatal CD-1 mouse model. Using non-linear statistical techniques, Gyürék and co-workers estimated that the 90% confidence range on the model predictions was approximately ± 0.7 log-units over the range of conditions investigated.

For each laboratory trial, the dissolved ozone concentration at the inlet of the reactive flow segment, C_0, and the first-order rate constant for ozone decay within the reactive flow segment, k_d, were determined by regression of Equation 2.10 to the measured ozone concentration profile using least-squares minimization criteria. The average mean residence time, t_m, previously measured in tracer tests (Figure 2.10 and Table 2.4), was used as the time variables in this analysis. For plug flow analysis, the estimated values of C_0, k_d and t_m were substituted directly into Equation 2.23. For segregated flow analysis, the batch kinetic expression for *Cryprosporidium parvum* oocyst inactivation by ozone, Equation 2.23, was substituted into Equation 2.20 to give the following expression:

$$\frac{N}{N_0} = \int_{t=0}^{t=\infty} anti\log\left(\frac{-mk'C_0{}^n}{(nk_d)^m}\gamma(m, nk_d t)\right)E(t)dt \tag{2.24}$$

Equation 2.24 was solved numerically on a Microsoft Excel 2000 computer spreadsheet using the trapezoidal rule with a time step of 1 s. Segregated flow analysis is inexact for non-first order reacting systems and yields an upper bound on conversion for those systems with reaction orders greater than one. The difference between estimated and real conversion becomes negligible for most reacting systems as the residence time distribution behavior approaches that of plug flow (Levenspiel 1972; Nauman and Buffham 1983).

Oocyst inactivation within the gas/liquid separator was estimated by assuming perfect plug flow and a constant ozone concentration. The geometric mean ozone concentration in the gas/liquid separator, C_{avg}, was estimated using Equation 2.12 and the average mean residence time, t_m, was determined from tracer test results (Figure 2.10 and Table 2.4). The estimated level of inactivation in the gas/liquid separator was calculated using Equation 2.22. This was then added to the inactivation predicted within the reactive flow segment by either plug flow or segregated flow analysis to yield the total estimated inactivation for the contactor. Because the inactivation curve of *Cryptosporidium parvum* predicted by the batch kinetics was non-linear, this additive correction was not exact. The magnitude of the error introduced, however, was expected to be minor because inactivation within the gas/liquid separator was expected to be small.

Laboratory Results

Ozonation conditions for each of the oocyst challenge experiments were summarized in Table 2.10. Corresponding values for the measured and the model-predicted *Cryptosporidium parvum* oocyst log inactivation for each laboratory condition were provided in Table 2.11. To avoid introduction of bias into the contactor modeling, the information in Table 2.10 was based on measurement of the dissolved ozone by the UV absorbance at 260 nm rather than by the indigo trisulphonate method. The reason for this choice of analytical methods was that the kinetic model of *Cryptosporidium parvum* inactivation by ozone used in the contactor analysis (Equation 2.23) was derived from measurements of dissolved ozone by UV absorbance at 260 nm (Gyürék et al. 1999).

In the batch reactor trials with *Bacillus subtilis* spores, which were carried out in ozone demand-free buffered laboratory water, the ratio of ozone concentration determined by the indigo method to the ozone concentration determined by UV absorbance was found to range from 0.95 to 1.35 with a mean of 1.17. In this relatively clean system, the indigo method, therefore, produced ozone concentrations that were about 20% higher than the UV absorbance method. This offset was most likely due to assumptions regarding the molar absorption coefficients used in the two methods. A wide range of UV molar absorption coefficients have been reported in the literature for aqueous ozone (Gordon et al. 1988). A recent study reported that the molar absorption coefficient for determination of the indigo trisulphonate method may vary by up to 25% depending on the source of the indigo reagent (Gordon et al. 2000). Using the baseline correction for the UV absorbance measurement (Equation 2.4), the ozone concentration determined by the indigo method in the *Cryptosporidium parvum* challenge experiments was found to be, on average, 1.27 times greater (range 1.07 to 1.55) than that measured by UV absorbance.

In trials 43 through 47, the average measured oocyst inactivation in the feed and static mixer outlet streams was 0.06 log-units and 0.19 log-units, respectively (columns 5 and 6 of Table 2.11). The average difference in inactivation between the feed and the static mixer outlet streams was, therefore, only 0.13 log-units. The distributions of these measured values were illustrated in the box diagram of Figure 2.20. The overlap of the 25th percentile regions with the

Table 2.10
Summary of ozonation conditions for the *Cryptosporidium parvum* challenge experiments in the laboratory study static mixer contactor at 22°C

				Gas/liquid separator			Reactive Flow Segment				Total
Trial	v_s m/s	*G/L* %	*TE** %	C_{avg}† mg/L	t_m min	$C_{avg}t_m$ mg/L× min	C_0† mg/L	t_m min	k_d min^{-1}	$C_{avg}t_m$ mg/L× min	$C_{avg}t_m$ mg/L× min
43	0.63	1.26	74	0.60	0.73	0.44	0.50	15	0.055	5.08	5.5
44	0.64	1.23	74	0.56	0.71	0.40	0.53	15	0.037	6.12	6.5
45	0.68	1.18	76	0.78	0.67	0.52	0.73	7.4	0.047	4.57	5.1
46	0.64	1.24	74	0.62	0.71	0.44	0.57	7.4	0.059	3.45	3.9
47	0.65	1.22	75	0.66	0.70	0.46	0.62	15	0.055	6.35	6.8
48	1.12	1.57	84	0.59	0.55	0.32	0.57	3.7	0.028	1.96	2.3
49	0.64	1.24	74	0.60	0.71	0.43	0.54	3.6	0.070	1.72	2.2
50	0.64	1.24	74	0.58	0.71	0.41	0.55	3.7	0.091	1.66	2.1
51	1.38	1.56	86	0.56	0.45	0.25	0.51	3.6	0.075	1.59	1.9
Control				0	0.7	0	0	15			

* Ozone transfer efficiency estimated from an empirical model
† Ozone concentrations were based on measurements by UV absorbance at 260 nm

Table 2.11
Measured and model-predicted *Cryptosporidium parvum* oocyst inactivation in the laboratory study static mixer ozone contactor at 22°C

	Model Predicted			Measured		
	Gas/liquid separator	Total at RFS Outlet				
Trial	Outlet	PFA	SFA	Feed	Static Mixer Outlet	RFS Outlet
43	0.4	2.8	2.7	-0.2	-0.3	3.1
44	0.4	3.1	3.0	-0.2	-0.2	> 3.3
45	0.4	2.6	2.5	-0.5	0.3	3.1
46	0.4	2.1	2.0	-0.7	1.0	2.3
47	0.4	3.2	3.1	0.5	0.1	3.4
48	0.3	1.4	1.4	0.6	ND	1.5
49	0.4	1.4	1.4	0.2	ND	1.7
50	0.4	1.4	1.4	0.2	ND	2.0
51	0.3	1.3	1.2	0.1	ND	1.5
Control				0.5	ND	0.4

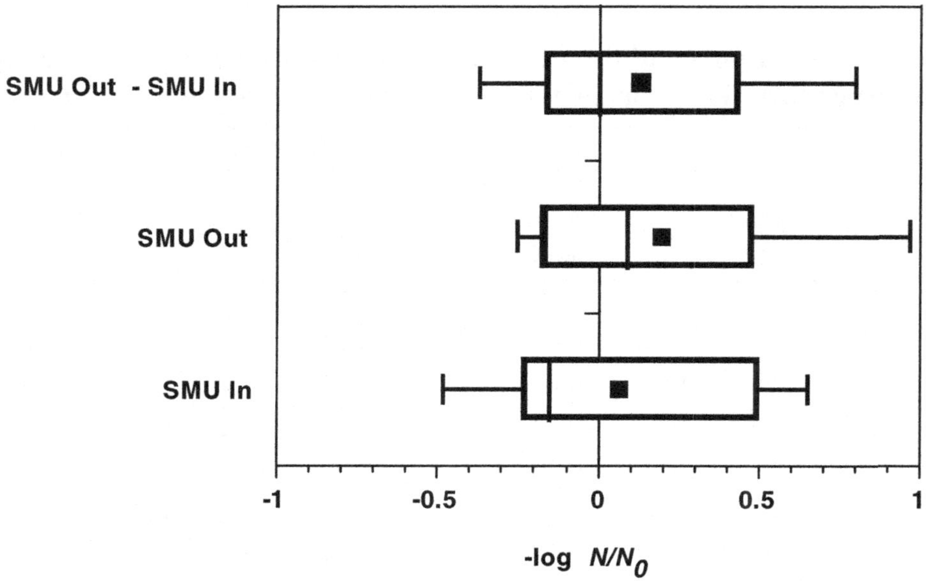

Note: From left to right, lines represent 10%, 25%, median, 75% and 90% percentile boundaries, respectively. The square symbol represents the mean.

Figure 2.20 Box plot of measured *Cryptosporidium parvum* inactivation at the inlet and outlet of the static mixer in the laboratory study

origin in Figure 2.20 was an indication that none of the three average outcomes differed significantly from zero. In addition, a two-tailed paired Student t-test showed that the difference in inactivation between the feed and static mixer outlet was not statistically significant at the 95% confidence level ($p = 0.55$). It was concluded, therefore, that the loss of oocyst infectivity due to laboratory factors other than experimental exposure to ozone was minimal. It was also concluded with some confidence that no significant inactivation of oocysts occurred within static mixer itself.

Measured oocyst inactivation was plotted as a function of the total dissolved ozone $C_{avg}t_m$ product in the static mixer contactor in Figure 2.21. The plotted values at the x-axis origin represent inactivation measured at the outlet of the static mixer where, because of the extremely short residence time, the $C_{avg}t_m$ product was essentially zero. Using non-linear least squares regression, a polynomial equation was fit to the inactivation curve of Figure 2.21. The best-fit value of the $C_{avg}t_m$ exponent of 0.6 indicates a non-linear relationship between oocyst inactivation and $C_{avg}t_m$. Furthermore, the exponent value was close to the values of the n (0.70) and m (0.73) exponent parameters of the Hom kinetic model of *Cryptosporidium parvum* inactivation by ozone determined by Gyürék and others (1999). The outcomes of two trials that were conducted at a higher superficial velocity, v_s, in the static mixer (trials 48 and 41) did not differ markedly from the outcomes of the remaining trials that were conducted at the lower v_s. This is an indication that oocyst inactivation was determined exclusively by exposure to dissolved ozone and was independent of the hydrodynamic conditions within the static mixer.

According to the model predictions of Table 2.11, most of the oocyst inactivation in the static mixer contactor was expected to take place during contact with dissolved ozone within the reactive flow segment. Less than 0.4 log-unit inactivation was attributed to contact within the gas/liquid separator. Also according to the model predictions, the residence time distribution in the

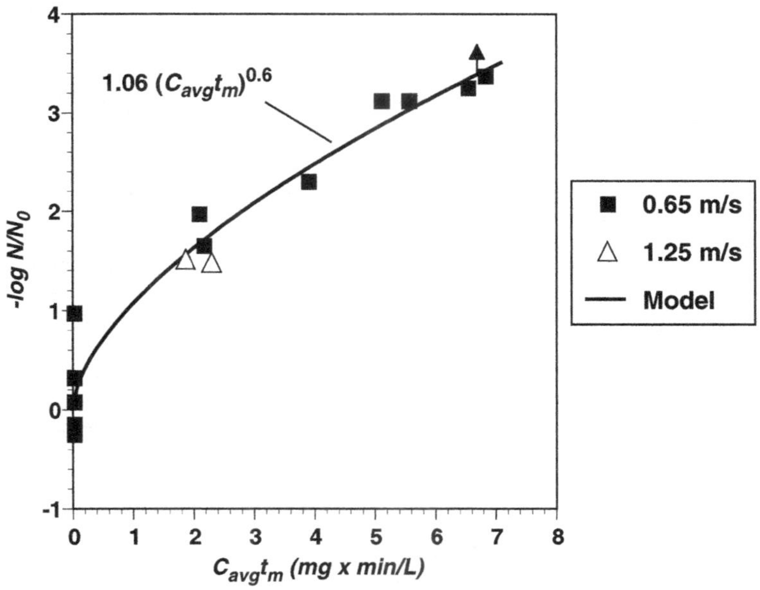

Note: The arrow indicates that the measured inactivation was beyond the level of detection of the infectivity analysis for this datum.

Figure 2.21 Measured inactivation of *Cryptosporidium parvum* in the laboratory study static mixer ozone contactor

reactive flow segment had little influence on oocyst inactivation. The difference between predictions of oocyst inactivation based on the assumption of perfect plug flow and those generated using segregated flow analysis and the measured exit age distribution was less than 0.1 log-unit (Table 2.11). Based on these modeling outcomes, the residence time behavior in the reactive flow segment, and in the contactor overall, was expected to result in *Cryptosporidium parvum* oocyst inactivation that was close to that of an ideal plug flow reactor.

Oocyst inactivation at the outlet of the reactive flow segment measured in the challenge experiments confirmed the model-predicted performance expectations. In fact, measured inactivation, was found to consistently exceed that predicted by the theoretical maximum predicted by the plug flow model by, on average, 0.3 log-units (Figure 2.22). For individual measurements, this difference was well within the approximate error bounds of the model (± 0.7 log-units). A two-tailed Student t-test on the average difference of all nine trials, however, revealed that the difference was statistically significant at the 95% confidence level ($p = 0.0007$).

One potential explanation for the observed difference between the measured and model-predicted inactivation was that the kinetic model of Gyürék and co-workers (1999) did not provide accurate predictions of the inactivation kinetics of the laboratory batch of oocysts in the laboratory water matrix. To test the kinetic model, batch reactor trials were done on the same days as the static mixer contactor experiments using the same oocysts suspended in the same water matrix. In 6 out of the 7 batch reactor trials, measured inactivation was found to be within the approximate error bounds (± 0.7 log-units) of the batch kinetic model (Figure 2.23). Moreover, the average difference between the model and the measured inactivation was only 0.04 log-units. A two-tailed pair Student t-test showed that this difference was not significant at the 95% confidence level ($p = 0.85$). The kinetic model of Gyürék and co-workers (1999), therefore, provided accurate and unbiased predictions of inactivation for the laboratory oocyst batch and the

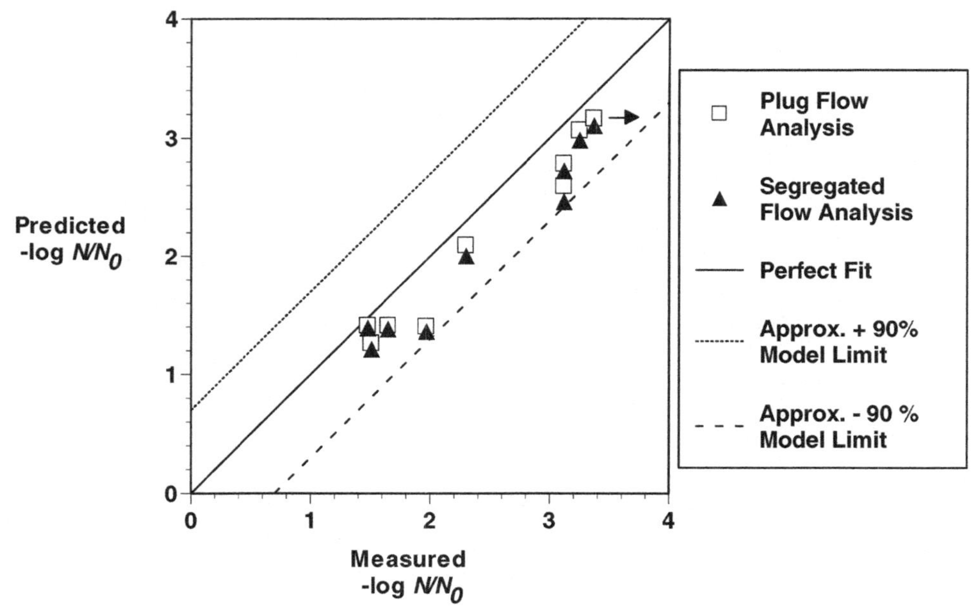

Note: The arrow indicates that the measured inactivation was beyond the level of detection of the infectivity analysis for this datum.

Figure 2.22 Comparison of model predicted and measured inactivation at the outlet of the reactive flow segment for each experimental trial with *Cryptosporidium parvum* oocysts in the laboratory study static mixer contactor at 22°C

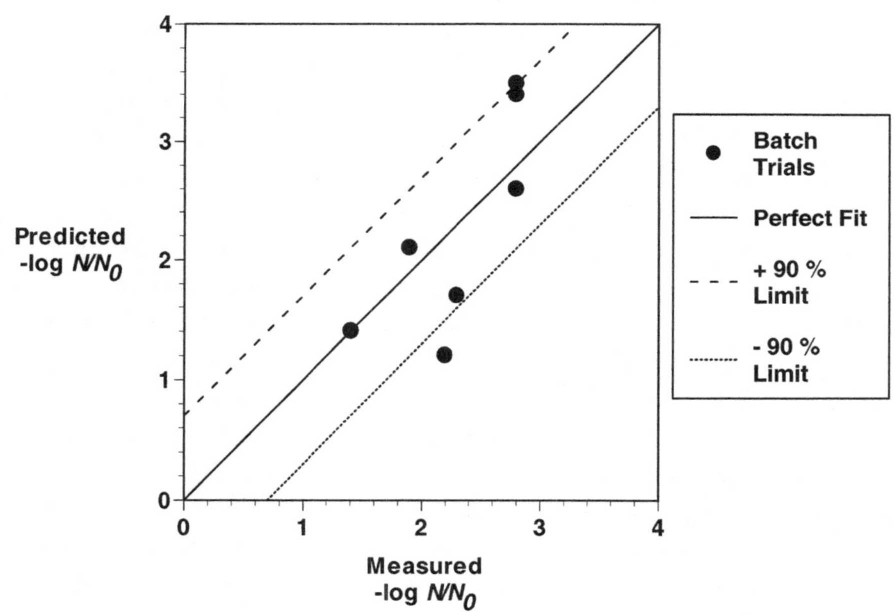

Note: The center diagonal line represents a perfect fit to the kinetic model of Gyürék and co-workers (1999).

Figure 2.23 Comparison of model-predicted and measured inactivation of the laboratory *Cryptosporidium parvum* oocysts suspended in the laboratory water matrix as determined in batch reactors at 22°C

water matrix. These variables were ruled out as potential explanations for differences between measured and model-predicted inactivation in the static mixer contactor.

The difference between the measured and model-predicted inactivation of 0.3 log units may have been evidence of a static mixer effect on oocyst inactivation. Given the relatively small magnitude of the observed difference, however, any claims of enhancement of ozone inactivation kinetics by virtue of initial contact within the static mixer must be viewed cautiously. The apparent enhancement effect may also have been explained by unidentified sources of systematic error. For example, a consistent bias in determination of the laboratory $C_{avg}t_m$ values in the static mixer contactor of between 2 and 6% was sufficient to explain the 0.3 log-unit offset.

Inactivation of *Giardia muris* in the Static Mixer Contactor

Laboratory Plan

A total of six challenge experiments were completed with *Giardia muris.* For each, the short gas/liquid separator configuration (H = 0.45 m) was employed and the feed water temperature was regulated to 22°C. Experiments were carried at nominal static mixer superficial velocities, v_s, of either 0.5 m/s or 1.4 m/s while maintaining approximately constant dissolved ozone conditions. *Giardia muris* oocysts were considerably more sensitive to the effects of ozone than *Cryptosporidium parvum* oocysts. Much lower target ozone exposures were required for challenge experiments with *Giardia muris* oocysts, compared to those used in the *Cryptosporidium parvum* oocyst challenges, to ensure that cyst inactivation remained within measurable limits. The length of the reactive flow segment was reduced to provide a maximum total contact time of approximately 3.7 minutes, and the target ozone concentration range was decreased from about 0.5 to 0.7 mg/L to 0.1 mg/L. In each experiment, samples of water were collected from the feed (after cyst addition but prior to ozone addition), the static mixer outlet and the reactive flow segment outlet for cyst recovery and infectivity analysis with the C3H/HeN mouse latent period model.

Contactor Modeling

For *Giardia muris*, the entire static mixer contactor was treated as a single ideal plug flow unit with a constant average ozone concentration. A simplified Hom model was used to describe the *Giardia muris* cyst inactivation kinetics by ozone. The kinetic model was developed using information from batch reactor ozonation experiments with *Giardia muris* that was extracted from the doctoral thesis of Charles Labatiuk (1992). The Labatiuk study was selected mainly because the same strain of *Giardia muris* cysts was used, and inactivation was assessed using a similar infectivity assay based on latent period of infection in C3H/He mice. In addition, Labatiuk used batch ozonation procedures that were very similar to those described in this study, including measurement of dissolved ozone by UV absorbance at 260 nm. Relevant information regarding the trials selected for model development are summarized in Table A.2 of Appendix A. These particular set of laboratory trials were selected from the Labatiuk thesis for examination and modeling because they were carried out at 22°C in phosphate buffered water and at near neutral pH.

The Hom kinetic model of Equation 2.23 was used to model the Labatiuk batch reactor data set with the additional simplifying assumption that the exponent parameters n and m have equal value. Inactivation thus became a non-first order function of the simple Ct product, that is, a

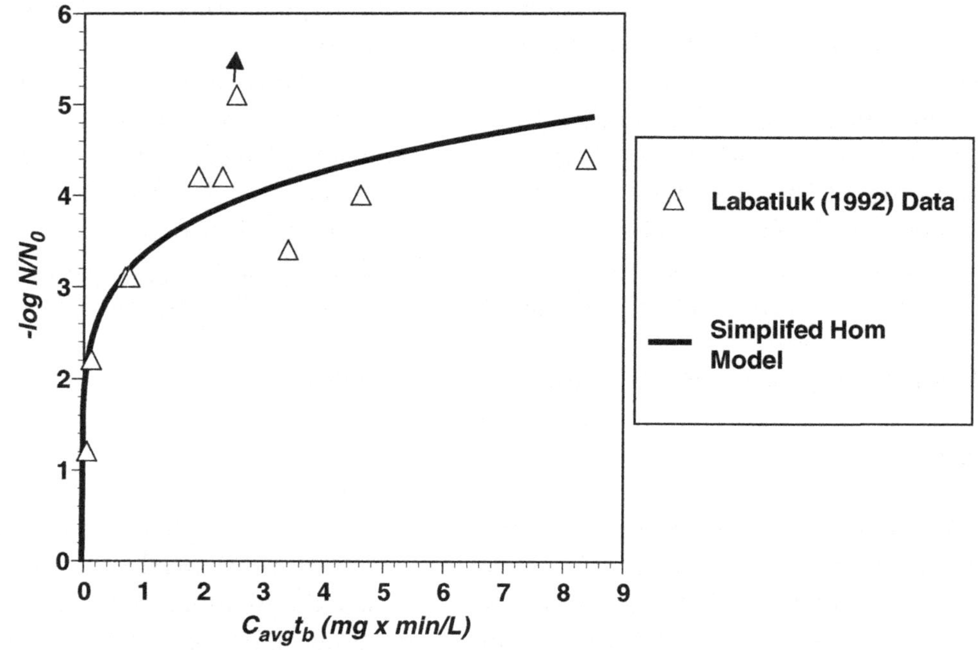

Note: Arrow indicates experimental results were beyond detection limit of infectivity assay.

Figure 2.24 Inactivation of *Giardia muris* by ozone in batch reactors as determined by latent period infectivity analysis in C3H/HeN mice (*Source:* Labatiuk 1992)

function of $(Ct)^n$. Using non-linear regression with least-squares minimization criteria, the best fit parameters of Equation 2.23 were determined to be $k' = 3.3$ and $n = m = 0.18$. The fit of the model to the Labatiuk dataset is shown in Figure 2.24.

Experimental Results

Ozonation conditions for each of the challenge experiments with *Giardia muris* are summarized in Table 2.12. The ozone conditions selected for analysis were those based on measurement of dissolved ozone by the UV absorbance method. The justification for this decision was similar to that described for analysis of the *Cryptosporidium parvum* challenge experiments. Cyst inactivation measured using the C3H/HeN mouse latent period model was summarized for each trial in Table 2.13. Model-predicted inactivation determined for the laboratory ozonation conditions is also provided in Table 2.13. Because of the sensitivity of the parasite to ozone and the non-linear inactivation kinetics (Figure 2.24), most of the total cyst inactivation within the static mixer contactor was predicted to occur within the gas/liquid separator. For this reason, the two-part contactor modeling approach used with *Cryptosporidium parvum*, would have generated serious errors for the case of *Giardia muris*. The simpler modeling approach described above was, therefore, adopted.

A box diagram of cyst inactivation measured in the feed and outlet of the static mixer is shown in Figure 2.25. Contrary to expectations, a slight increase in infectivity (negative inactivation) in the cysts in both the feed and static mixer outlet was observed. Average inactivation levels were -0.2 log-units and -0.4 log-units in the feed and static mixer outlet,

Table 2.12
Summary of ozonation conditions for the *Giardia muris* challenge experiments on the laboratory study static mixer contactor

				Gas/liquid separator			Reactive Flow Segment				Total
Trial	v_s m/s	*G/L* %	*TE** %	C_{avg}† mg/L	t_m min	$C_{avg}t_m$ mg/L× min	C_0† mg/L	t_m min	k_d min^{-1}	$C_{avg}t_m$ mg/L× min	$C_{avg}t_m$ mg/L× min
54	0.46	1.70	71	0.11	1.21	0.13	0.08	3.7	0.15	0.23	0.37
55	1.15	1.50	86	0.08	0.60	0.05	0.07	3.6	0.18	0.20	0.25
56	0.39	1.98	68	0.13	1.31	0.17	0.11	3.7	0.10	0.33	0.51
57	1.32	1.30	89	0.10	0.51	0.05	0.09	3.6	0.13	0.26	0.31
58	0.43	1.80	69	0.09	1.25	0.11	0.07	3.6	0.23	0.18	0.28
59	1.33	1.36	88	0.08	0.51	0.04	0.06	3.7	0.20	0.16	0.20

* Ozone transfer efficiency estimated from an empirical model
† Ozone concentrations were based on measurements by UV absorbance at 260 nm

Table 2.13
Measured and model-predicted *Giardia muris* cyst inactivation in the laboratory-scale static mixer ozone contactor

	PFA Model Predicted		Measured		
Trial	Gas/liquid separator Outlet*	RFS Outlet*	Feed*	Static Mixer Outlet*	RFS Outlet*
54	2.3	2.8	0.18	-0.22	-0.23
55	2.0	2.6	-0.22	-0.32	> 2.20
56	2.4	3.0	-0.36	-0.11	1.5
57	2.0	2.7	0.39	-0.73	1.3
58	2.3	2.7	-0.5	-0.04	> 3.0
59	1.9	2.5	-0.49	-0.72	> 3.4
Control			-0.19	-0.20	0.63

* Values are in log-units

respectively. This may have reflected a slight increase in the infectivity of the cyst preparations used in experiments relative to those used to develop the latent period dose-response model (Figure 2.7). The average difference between inactivation in the feed and the static mixer outlet was -0.1 log units. Despite the relative sensitivity of these microorganisms to ozone, no *Giardia muris* cyst inactivation was attributed to direct ozone contact within the static mixer itself.

Giardia muris cyst inactivation is plotted as a function of the measured dissolved ozone $C_{avg}t_m$ product in static mixer contactor in Figure 2.26. Predictions of the simplified Hom model are also plotted in Figure 2.26 for comparison purposes. Precise interpretation of the results was made difficult by the degree of variability in the experimental outcomes. One of the experimental

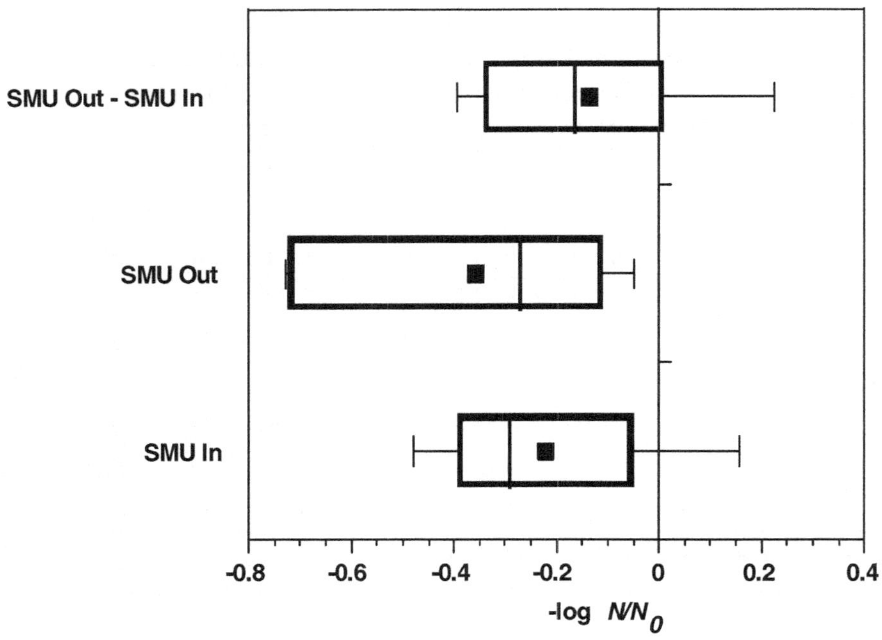

Note: From left to right, lines represent 90%, 25%, median, 25% and 90% percentile boundaries, respectively. The square symbol represents the mean.

Figure 2.25 Box plot of measured *Giardia muris* inactivation at the inlet at outlet of the static mixer in the laboratory study

difficulties encountered in the *Giardia muris* challenge experiments was the relatively low recovery of cysts from the outlet of the reactive flow segment. Measured cyst recoveries ranged from 0.2 to 19%. In the experiment in which cyst recovery was only in 0.2% (Trial 54), essentially no cyst inactivation was observed at the outlet of the reactive flow segment. Such low recoveries reduced the confidence in the laboratory measurement of cyst inactivation because those cysts that were not recovered may have been mostly inactivated by ozone. In comparison, recoveries in the *Cryptosporidium parvum* trials were consistently higher, at between 6 and 67%.

A second difficulty of the *Giardia muris* analysis was posed by the fact that in three of the trials (Trials 55, 58 and 59) the measured inactivation was beyond the limits of detection of the infectivity assay. As an approximate and conservative analysis, average inactivation of all trials was estimated by setting the outcomes of these trials at the detection limit. Because of the extremely low recovery, the result of Trial 54 was excluded. The estimated average inactivation was 2.3 log units. For the range of laboratory $C_{avg}t_m$ products (approximately 0.2 to 0.5 mg × min/L), the simplified Hom model predicted between 2.5 log-units and 2.9 log-units inactivation. The average measured inactivation, therefore, was not considerably different from that predicted using the simplified Hom model and assuming perfect plug flow. The result also agrees well with the ozone resistance of *Giardia muris* cysts reported by others in the literature. For example, the Ct product required for 2 log-unit inactivation of *Giardia muris* at 25°C and pH 7 in batch reactors, was estimated to be 0.27 mg × min/L (Wickramanayake et al. 1984).

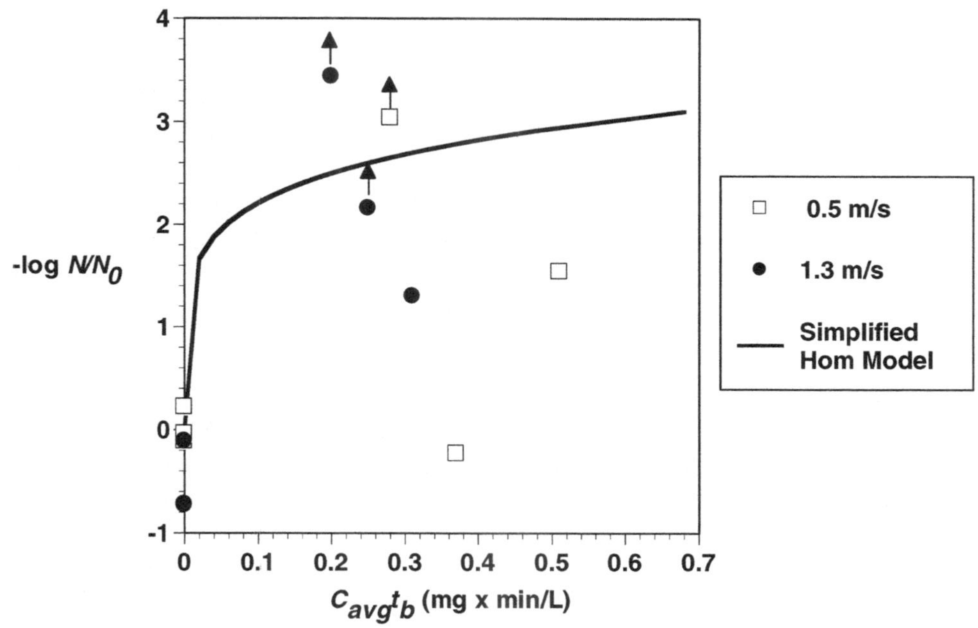

Note: The solid line represents predictions of the simplified Hom batch kinetic model. Arrows indicate outcomes were beyond the detection limit of the infectivity assay.

Figure 2.26 Inactivation of *Giardia muris* in the laboratory study static mixer ozone contactor at different superficial velocities, v_s, in the static mixer

Kinetics of *Bacillus subtilis* Inactivation and Modeling

Experimental Plan

The kinetics of inactivation of the cultured strain *Bacillus subtilis* (ATCC 6633) spores used in the laboratory and pilot-scale studies were determined in batch reactor ozonation experiments as explained earlier. Experiments were carried out in ODF 0.05 M phosphate buffered laboratory water at two pH levels (6 and 8) and at three temperatures (3°C, 12°C and 22°C). A minimum of eight batch trials, spanning a range of dissolved ozone concentrations and contact times, were completed at each pH and temperature combination. A total of fifty batch ozonation trials were conducted in random order over the course of an 8-week period using a single stock preparation of spores (preparation A). Randomization reduced the impact of changes in the stock preparation with time on subsequent kinetic modeling. Both UV absorbance at 260 nm and the indigo trisulphonate methods were used to measure the dissolved ozone concentration, however, only the measurements based on UV absorbance were used for the analysis that follows. This allowed a direct comparison between spore inactivation kinetics and a kinetic model of *Cryptosporidium parvum* oocyst inactivation by ozone. The *Cryptosporidium parvum* model was developed using similar batch reactors protocols and ozone measurements based on UV absorbance (Gyürék et al. 1999; Li et al. 2001).

Kinetic Modeling

A kinetic model was developed to describe the ozone inactivation of *Bacillus subtilis* spores measured in the batch reactor experiments. The main purpose of the model was to facilitate a comparison between the laboratory determined ozone inactivation kinetics of the spores, and the ozone inactivation kinetics of *Cryptosporidium parvum* oocysts previously determined by other researchers (Li et al. 2001), especially at temperatures other than the one studied in this project. A simple first-order, Chick-Watson rate law is clearly not adequate to describe the non-linear characteristic of the observed spore inactivation curves. The presence of an initial shoulder in an inactivation curve is an indication of inactivation process that arises from cumulative damage and that can be described by a multi-target model (Hiatt 1964). Tailing is an indication of the heterogeneity in the spore population and the presence of a resistant sub-population of microorganisms (Cerf 1977). The following two-term rate model was, therefore, used to describe the observed spore inactivation:

$$N = N_{0,b}[1-(1-e^{-k_b Ct})^{n_c}] + N_{0,r}e^{-k_r Ct} \tag{2.25}$$

$$f = \frac{N_{0,b}}{N_{0,b} + N_{0,r}} \tag{2.26}$$

The first term of Equation 2.25 represented inactivation of free spores according to the multi-target model. The second term represented inactivation of resistant spores by a first-order process. The variables were defined as follows: N was the total number of live colony forming entities after a given dissolved ozone Ct exposure, $N_{0,b}$ and $N_{0,r}$ were the initial number of normal spores and resistant spores, respectively, n_c was the number of critical targets that must be inactivated in order to result in the death of a normal spore, and k_b and k_r were rate constants. The parameter f, defined in Equation 2.26, specified the initial fraction of resistant spore entities in the total spore population (In this treatment, a spore agglomerate was equivalent to a single spore).

To account for the temperature dependence of the spore inactivation kinetics, the value of the rate constants, k_b and k_r, were assumed to follow a van't Hoff-Arrhenius temperature dependence according to:

$$\frac{k_1}{k_2} = \exp\left[-\frac{E_A}{R}\left(\frac{1}{T_1} - \frac{1}{T_2}\right)\right] \tag{2.27}$$

where E_A is the activation energy of the reaction, R is the universal gas constant and T_1 and T_2 are the respective absolute temperatures. Equation 2.27 can be re-written as:

$$\frac{k_1}{k_2} = \theta^{T_1 - T_2} \tag{2.28}$$

where the parameter, θ, is given by exp (E_A/RT_1T_2). Over a sufficiently small temperature range, θ is approximately constant. For spore kinetic modeling, the temperature dependence of the two rate constants in Equation 2.25 were mathematically described as follows:

$$k_{b,T} = k_{b,12}\theta_b^{T_1-12} \tag{2.29}$$

$$k_{r,T} = k_{r,12}\theta_r^{T_1-12} \tag{2.30}$$

The reference temperature of 12°C was the approximate midpoint of the experimental range of 3 to 22°C.

The kinetic model, given by Equations 2.25, 2.26, 2.29 and 2.30 was fit to the laboratory inactivation curves using classical non-linear least-squares regression. The criteria for best fit of the model was minimization of the sum of the squares of the residuals described as follows:

$$\min \sum \left[\log\left(\frac{N}{N_0}\right)_{\text{model}} - \log\left(\frac{N}{N_0}\right)_{\text{measured}} \right]^2 \tag{2.31}$$

Calculations were done in a Microsoft Excel® 2000 spreadsheet using the Solver function with the Newton search method and the convergence tolerance set to 0.0001.

The two-term kinetic model was found to provide a good fit to the observed experimental outcomes at each of the three laboratory temperatures (Figures 2.27, 2.28 and 2.29). Best-fit model parameters are summarized in Table 2.14. The estimated values of the activation energy of

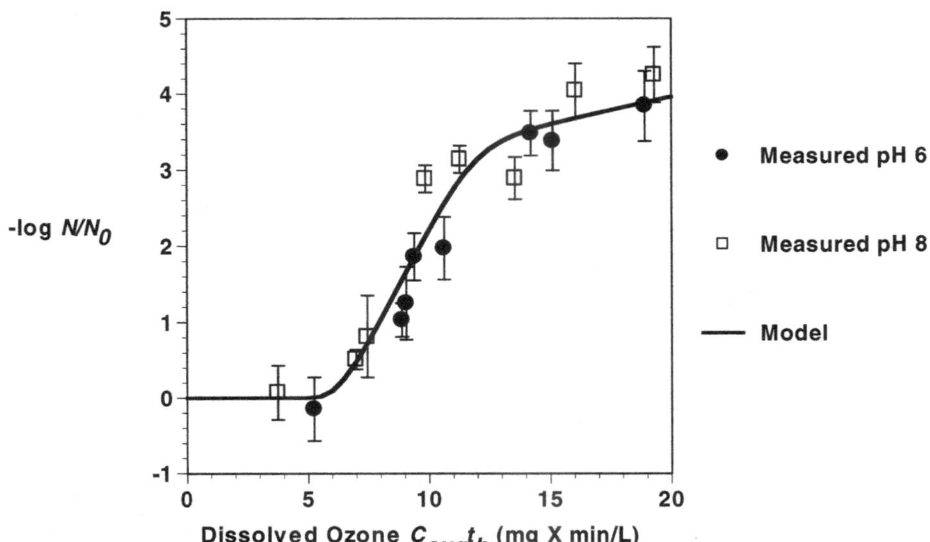

Note: Each datum represents the average enumeration of three samples extracted from each of a single reactor flask and a single control flask. The solid line represents the best-fit model predictions.

Figure 2.27 Ozone inactivation of *Bacillus subtilis* spores in ozone demand-free buffered laboratory water at 3°C

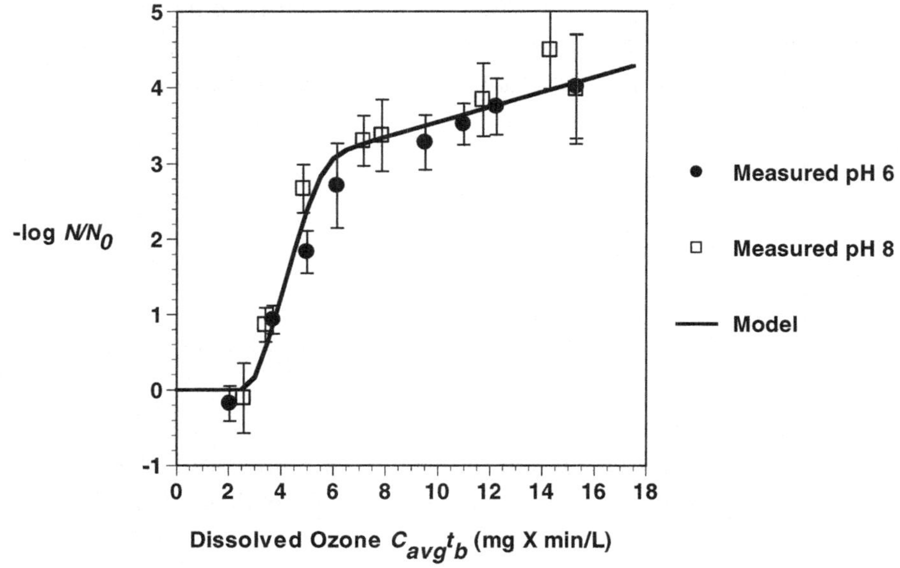

Note: Each datum represents the average enumeration of three samples extracted from each of a single reactor flask and a single control flask. The solid line represents the best-fit model predictions.

Figure 2.28 Ozone inactivation of *Bacillus subtilis* spores in ozone demand-free buffered laboratory water at 12°C

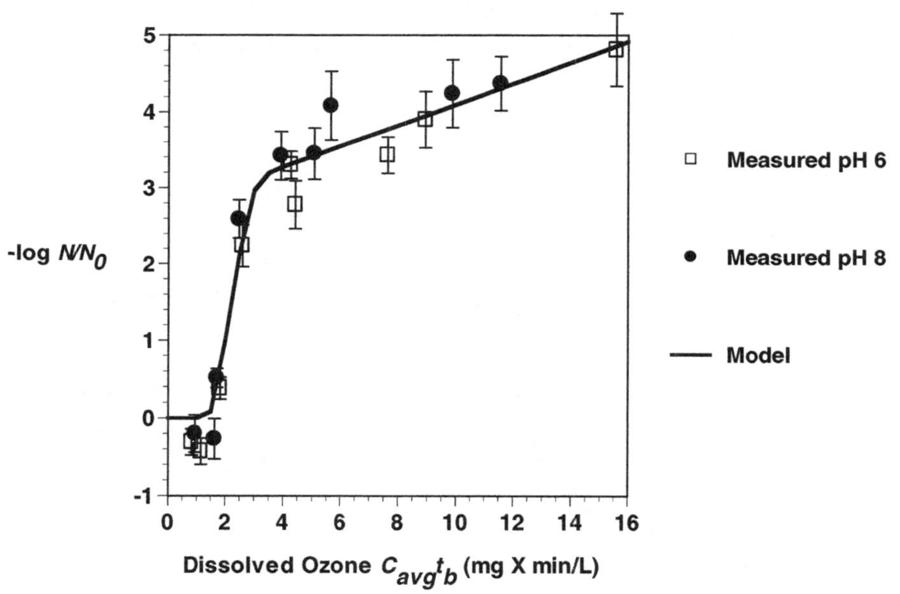

Note: Each datum represents the average enumeration of three samples extracted from each of a single reactor flask and a single control flask. The solid line represents the best-fit model predictions.

Figure 2.29 Ozone inactivation of *Bacillus subtilis* spores in ozone demand-free buffered laboratory water at 22°C

Table 2.14
Best-fit parameters of the *Bacillus subtilis* spore inactivation kinetic model

Parameter	Physical Significance	Best-Fit Value	Units
k_b	Rate constant for inactivation of normal spores	2.8	L mg^{-1} min^{-1}
θ_b	Temperature coefficient for inactivation of normal spores	1.08	none
k_r	Rate constant for inactivation of resistant spore units	0.23	L mg^{-1} min^{-1}
θ_r	Temperature coefficient for inactivation resistant spore units	1.04	none
n_c	Number of targets for inactivation of normal spores	8500	number
f	Fraction of resistant spore units in the total population	0.003	none
$E_{A,b}$	Estimated activation energy for normal spores	49	kJ mole^{-1}
$E_{A,r}$	Estimated activation energy for resistant spore units	28	kJ mole^{-1}

the kinetic rate constants, k_b and k_r were determined from the best-fit values of the temperature coefficients θ_b and θ_r. The value of 49 kJ/mole for the activation energy of the normal spores was very close to the activation energy reported for the rate constant of *Cryptosporidium parvum* inactivation by ozone of 51.7 kJ/mole (Li et al. 2001). This suggests that the temperature dependence of free spore inactivation was similar to that of *Cryptosporidium parvum*. The inactivation energy of the resistance spores, at 28 kJ/mole, was somewhat lower indicating a different rate controlling process.

Results

Spore inactivation in the batch reactors was a non-linear function of the average dissolved ozone concentration-time product, $C_{avg}t_b$, (Figures 2.27, 2.28 and 2.29). At each temperature, the inactivation curves exhibited the same characteristic non-linear sigmoidal shape, with a distinct shoulder region at low $C_{avg}t_b$, followed by a curvi-linear region and finally by a region of tailing at high $C_{avg}t_b$. An increase in temperature had the effect of reducing the size of the shoulder region and increasing the slope of the both the curvilinear and tailing regions. Outcomes of trials with similar $C_{avg}t_b$, but obtained with different combinations of ozone concentration and contact time (C_{avg} and t_b) were found to lie on the same inactivation curve. Inactivation curves measured at pH 6 and 8 did not appear to differ appreciably. These observations suggested that, within the range of laboratory conditions investigated (pH of 6 to 8 and C_{avg} of 0.5 mg/L to 1.7 mg/L), spore inactivation was determined mainly by temperature and Ct product. Other researchers reported that inactivation of *Cryptosporidium parvum* oocysts was essentially independent of pH between pH 6 and 8 (Gyürék et al. 1999; Li et al. 2001).

Tailing in the inactivation curves of bacterial spores and viruses exposed to various types of treatment is a frequently observed but often poorly understood phenomenon (Cerf 1977; Hiatt 1964). Other investigators have observed tailing in the survival curves of bacterial spores exposed to ozone but offered little explanation or discussion (Facile et al. 2000; Finch and Choe 1999). Though not proven, the presence of a sub-population of spores with a much greater resistance to ozone than the rest of the population was suspected as the cause of the tailing observed in this study. Microscopic observations revealed that the prepared spores tended to form agglomerates or "clumps" that were bound to extra-cellular material. Although the spore preparations were

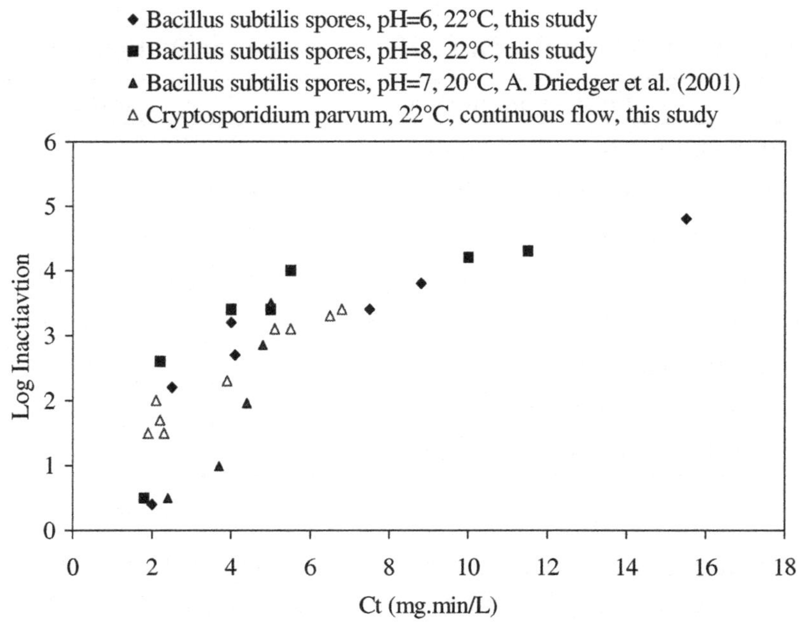

Figure 2.30 Comparison of *Bacillus subtilis* spore inactivation in batch experiments with *Cryptosporidium parvum* oocyst inactivation in the lab-scale ozone static mixer contactor at 22°C

vigorously mixed prior to use in the experiments to promote break-up of agglomerates, it is unlikely that agglomerates were completely eliminated. Spores near the interior of these agglomerates may have been protected from the effects of ozone in the bulk solution and, therefore, may have survived greater ozone exposures than free spores.

***Suitability of* Bacillus subtilis *as a Surrogate for* Cryptosporidium parvum**

Figure 2.30 compares the inactivation curves for *Bacillus subtilis* and *Cryptosporidium parvum,* using the available experimental data at 22°C from this work. A good agreement in inactivation of the two microorganisms at various Cts was observed. The results of batch trials with *Bacillus subtilis* spores conducted in tap water at 16°C are shown in Figure 2.31. Due to lack of experimental data for *Cryptosporidium parvum* inactivation at 16°C, the inactivation of *Bacillus subtilis* spores observed in this study is compared to inactivation of *Cryptosporidium parvum* oocysts predicted by an Incomplete gamma Hom kinetic model with temperature correction (Gyürék et al. 1999; Li et al. 2001). Predictions are based on the same experimental ozone conditions measured in this study. The developers of the *Cryptosporidium parvum* model used similar well-mixed, batch reactors in their experiments and conducted experiments at temperatures ranging from 1°C to 22°C. In general, resistance of spores and oocysts to ozone is comparable, although not identical at this temperature. The same Ct conditions are required for 2-log units inactivation of the two organisms. However, the overall shape of the inactivation curves differ somewhat. *Bacillus subtilis* spores are characterized by a more pronounced shoulder at low ozone Cts and greater tailing at high Cts. A recent study reported that *Bacillus subtilis* was a found to be a good indicator for *Cryptosporidium parvum* at 20 to 30°C, but at lower

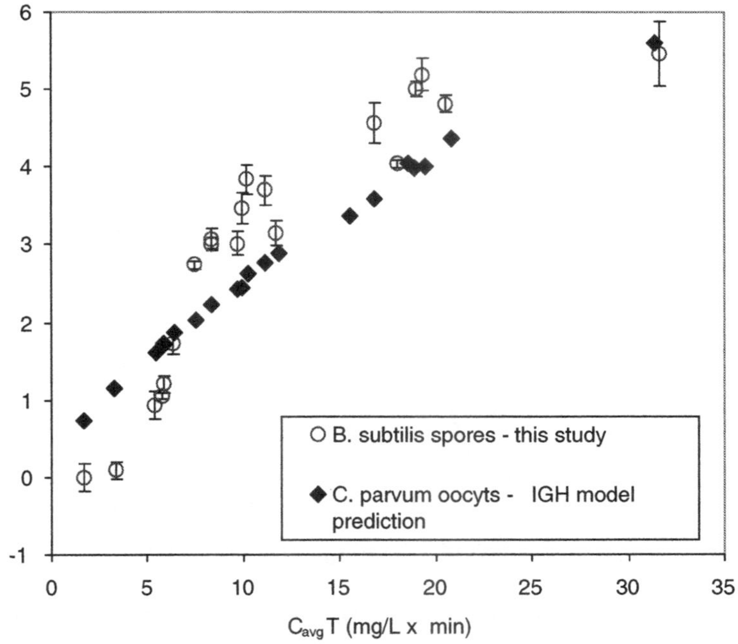

Figure 2.31 Comparison of *Bacillus subtilis* spore inactivation in batch experiments (16°C) with *Cryptosporidium parvum* oocyst inactivation predicted by an incomplete gamma Hom kinetic model (Gyürék et al. 1999; Li et al. 2001)

temperatures *Bacillus subtilis* was inactivated more readily than *Cryptosporidium parvum* (Driedger et al., 2001). It appears that the inactivation by ozone at temperatures ≥ 16°C is comparable for these two microorganisms. Hence, *Bacillus subtilis* can serve as a useful surrogate for *Cryptosporidium parvum* in disinfection studies.

Comparison of Inactivation Curves Based on Modeling

Comparisons of the model-predicted ozone inactivation curves for *Bacillus subtilis* spores and *Cryptosporidium parvum* oocysts for different temperatures are shown in Figures 2.32, 2.33 and 2.34. The *Cryptosporidium parvum* curves were calculated with the temperature-corrected Incomplete gamma Hom kinetic model developed by Li and co-workers (Li et al. 2001). This model was developed as part of a separate project sponsored by the Awwa Research Foundation (Project No. 348). These researchers used batch reactor protocols and ozone residual measurement methods identical to those used in this study. Differences in the inactivation curves were, therefore, attributed to differences between how ozone reacts with and inactivates spores and oocysts, respectively.

At low Ct, oocyst inactivation was greater than spore inactivation, mainly because of the shoulder region of the spore inactivation curve. At approximately 1 log-unit inactivation, the inactivation curves intersect and spore and oocyst inactivation requirements were approximately equal at all temperatures. As Ct was increased, spore and oocyst inactivation began to diverge rapidly. At the highest Ct levels, spore inactivation exceeded oocyst inactivation by as much as one to two log-units. The difference between spore and oocyst inactivation at high Ct was greatest at the lowest temperature (Figure 2.33). Based on this modeling, a 2-log inactivation of

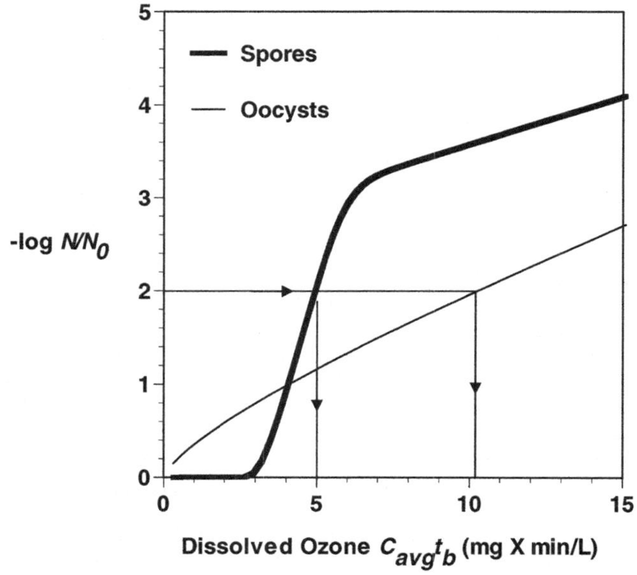

Note: Model simulations are based on ozonation conditions of $C_0 = 1.25$ mg/L, $k_d = 0.05$ mg/L.

Figure 2.32 Comparison of model-predicted ozone inactivation curves of *Bacillus subtilis* ATCC 6633 spores and *Cryptosporidium parvum* oocysts at 12°C and pH 6 to 8

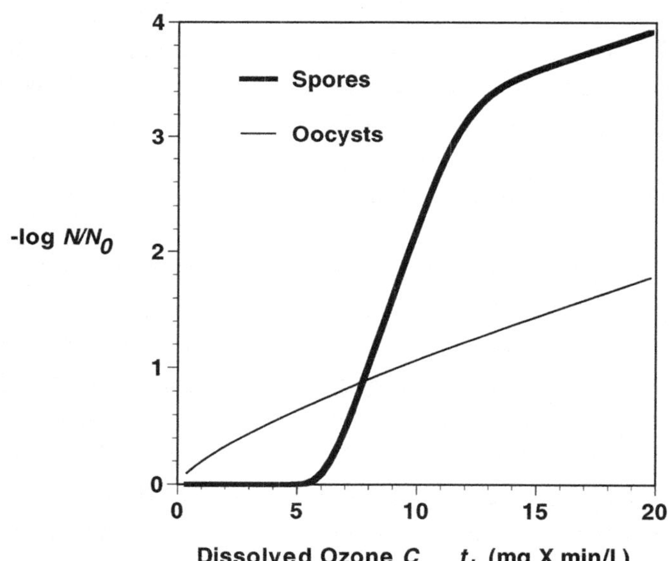

Note: Model simulations are based on the following ozonation conditions: $C_0 = 1.5$ mg/L, $k_d = 0.025$ mg/L.

Figure 2.33 Comparison of model-predicted ozone inactivation curves of *Bacillus subtilis* ATCC 6633 spores and *Cryptosporidium parvum* oocysts at 3°C and pH 6 to 8

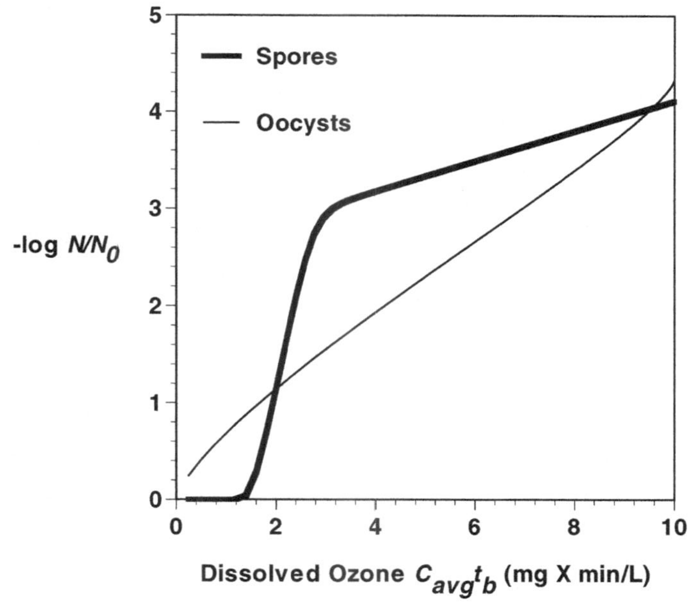

Note: Model simulations are based on ozonation conditions of $C_0 = 1.0$ mg/L, $k_d = 0.1$ mg/L.

Figure 2.34 Comparison of model-predicted ozone inactivation curves of *Bacillus subtilis* ATCC 6633 spores and *Cryptosporidium parvum* oocysts at 22°C and pH 6 to 8

Cryptosporidium parvum can be achieved at 2 to 2.3 times the equivalent *Bacillus subtilis* Ct value.

Clearly, more experimental inactivation data needs to be developed at cold water temperatures especially in continuous flow systems such as the one used in this study. This data needs to be compared to inactivation data obtained for *Bacillus subtilis* (also at cold temperatures) to develop the equivalent Ct value for inactivation of *Cryptosporidium parvum. Bacillus subtilis* spores are also useful for evaluation of the hydrodynamic efficiency of alternative ozone contactor designs.

By comparing the measured spore inactivation in an ozone contactor for a given ozonation condition to the spore inactivation predicted by batch reactor studies, hydrodynamic efficiency can be assessed. Problems such as flow short-circuiting and excessive back-mixing can be detected. In principle, the same information can be obtained using tracer tests. A spore assay, however, will provide a greater level of sensitivity than most tracer tests because spore concentrations can be measured with relative ease over a range or four orders of magnitude (or more). Secondly, spores may be better representatives of the hydrodynamic behaviour of particulate pathogenic microorganisms in contacting systems than the dissolved chemicals normally used in tracer tests.

Batch Reactor Ozonation Trials With Natural Waters

Experimental Plan

A series of batch ozonation trials were completed using *Bacillus subtilis* spores suspended in samples of natural waters shipped to the University of Alberta (UOA) from the pilot test sites during the last quarter. The purpose of these trials was to compare the spore inactivation performance of the pilot-scale static mixer unit at each test site relative to batch inactivation. Water samples were shipped to the UOA and stored at 4°C until use in experiments. All experiments were completed within 6 weeks of receiving the water sample at UOA.

Samples of raw water were collected during pilot-scale testing of the static mixer unit at three locations: Los Angeles Department of Water and Power (LADWP), the Elizabethtown Water Company (EWC) and the Chertsey water treatment plant, UK (CWTP). For each water sample, a set of batch ozonation experiments was conducted with *Bacillus subtilis* ATCC 6633 spores in well-stirred 250 mL Erlenmeyer flasks according to the procedure presented earlier. For the LADWP and EWC trials, a sample of the spore stock that was used in the pilot static mixer trials was shipped along with the water samples and this spore stock was used in the batch ozonation experiments. For the CWTP trials, spores generated at UOA were used in the batch ozonation experiments. Batch trials were conducted at various ozone doses and contact times to provide a range of dissolved ozone $C_{avg}t$ products that was comparable to that of the corresponding field trials. During each batch trial, the water temperature was maintained at the same temperature as the corresponding field trial by immersion of the batch reactor flasks in a thermostatically controlled water bath.

Results

When comparing spore inactivation measured in the static mixer contactor against inactivation in well-stirred batch reactors, possible sources of deviation include:

1. A direct effect of gas-liquid contacting in the static mixer on spore viability;
2. Effect of the hydraulics at the RTD in the reactive flow segment (i.e. contact columns) on the efficiency of spore inactivation;
3. Differences in the characteristics of the spore stock used;
4. Differences in the methods or materials used to measure dissolved ozone;
5. Differences in the methods used to enumerate spores; and
6. Effect of sample shipping or storage on spore viability.

Previous work with the laboratory-scale static mixer demonstrated that there was no appreciable effect of either item 1 or 2 on spore inactivation in the static mixer contactor. As discussed earlier, agreement between spore inactivation curves measured in the laboratory-scale static mixer contactor and in the batch reactors at UOA was very good, as items 3, 4, 5, and 6 noted above, were essentially eliminated as variables.

Figures 2.35, 2.36 and 2.37 show similar comparisons for the three pilot test sites investigated. For the EWC water (Figure 2.35), the result is similar to that of the UOA trials. There was no discernable difference between inactivation measured in the pilot-scale static mixer contactor and inactivation measured in the batch reactor. For the LADWP water, the inactivation curves diverged somewhat at the higher $C_{avg}t$ products but were still within 1 log-unit (Figure 2.36).

The measured spore inactivation in the pilot-scale static mixer greatly exceeded than measured in the batch reactor experiments for the CWTP (Figure 2.37). Note that for each of the comparisons discussed in this report, with the exception of the CWTP comparison, the same spore stock was used for the static mixer pilot and batch reactor trials. For the CWTP trials, spores generated at UOA were used in the batch ozonation experiments. For this reason, comparisons

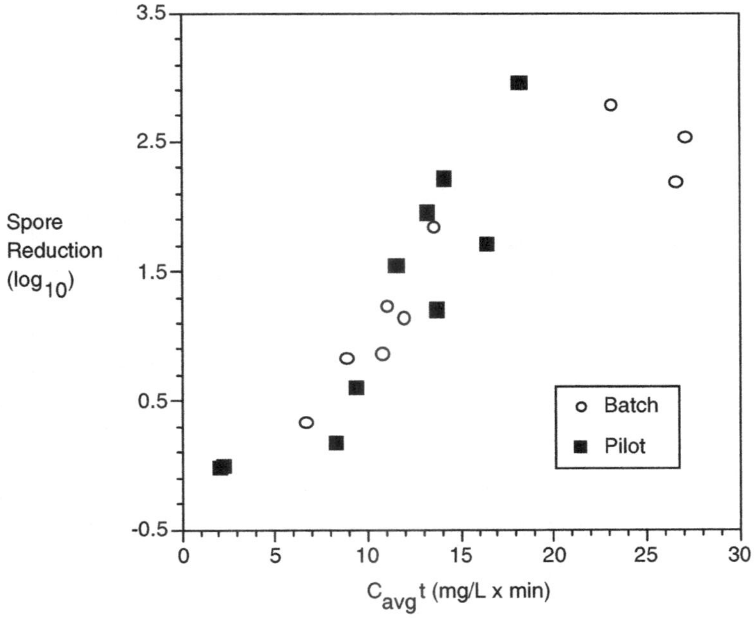

Figure 2.35 Comparison of *Bacillus subtilis* spore inactivation in the pilot-scale static mixer ozone contactor and in well-stirred batch ozone reactors using water from EWC at 5°C

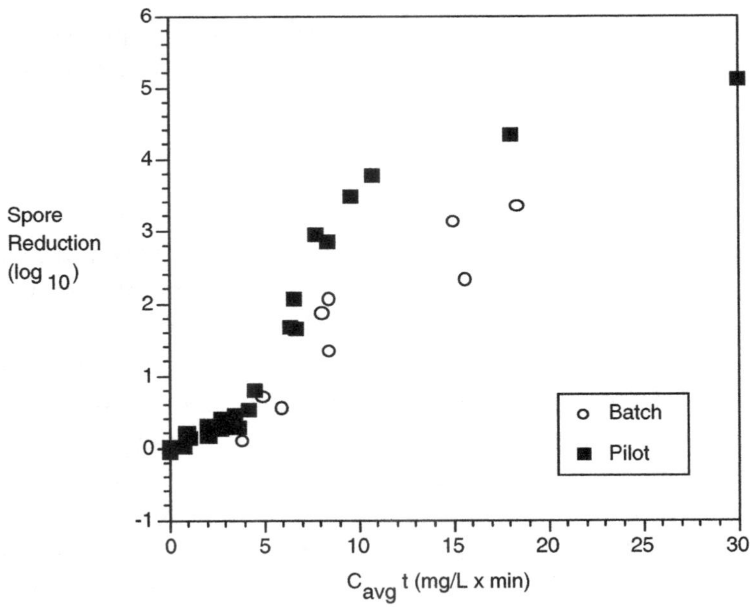

Figure 2.36 Comparison of *Bacillus subtilis* spore inactivation in the pilot-scale static mixer ozone contactor and in well-stirred batch ozone reactors using water from LADWP at 14°C

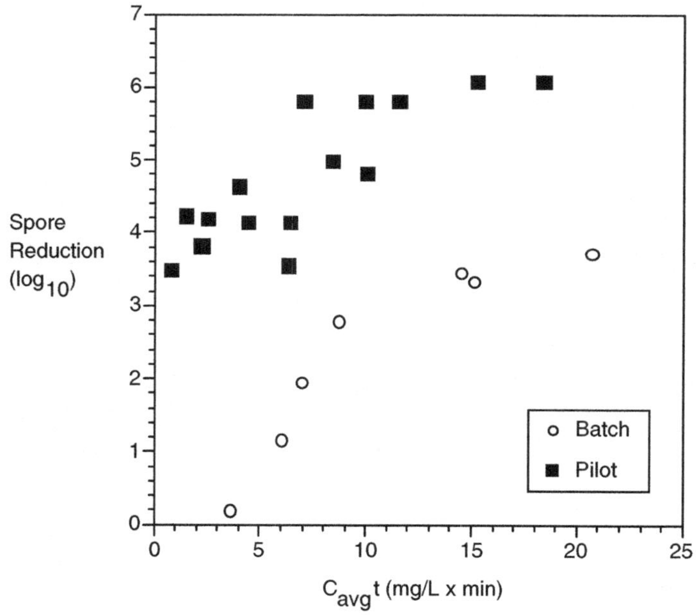

Figure 2.37 Comparison of *Bacillus subtilis* spore inactivation in the pilot-scale static mixer ozone contactor and in well-stirred batch ozone reactors using water from CWTP at 18°C

between spore inactivation in different contactors should ideally be made using the same spore stock solutions, of the same age, and enumerated at the same time with identical methods. These findings place emphasis on the need for strict protocols and good quality control measures when using cultured *Bacillus subtilis* spores to assess the microorganism inactivation performance of ozonation processes.

SUMMARY

Ozone transfer efficiency of the laboratory-scale unit was influenced by static mixer hydrodynamic parameters. A mathematical model was developed that related transfer efficiency to the superficial liquid velocity, v_s, and the gas-liquid flow rate ratio, *G/L*, within the static mixer. Transfer efficiency was greatest at maximum v_s and minimum *G/L*. This finding was not unexpected and was consistent with reports from previous studies on ozone dissolution in static mixers (Martin et al. 1992; Richards and Fleischman 1975; Zhu et al. 1989b). Improved transfer efficiency at low gas flows is an advantage of static mixers because it means that these devices can be effectively integrated with modern, high-concentration ozone generators. In comparison, conventional fine bubble diffusion systems are limited by a requirement of sufficient gas flow rate to ensure adequate gas-liquid mixing and ozone dissolution (Schulz et al. 1995). Modeling of the results showed that 94 % of the ozone transfer occurred in the static mixer itself and the gas-liquid separator provides little additional transfer. The static mixer was capable of achieving close to maximum theoretical ozone mass transfer. It was found that the outlet piping and the gas/liquid separator downstream of the static mixer play a significant role in determining overall ozone transfer efficiency. We believe that all the elements of the contacting system must be considered in the process design in order to properly optimize transfer efficiency.

The results showed that a level of disinfection equivalent to an ideal well-mixed batch system can be achieved in an ozone static mixer – plug flow type contactor system given the same average ozone concentration and contact time. Disinfection results obtained in the lab-scale contactor can used to predict disinfection in the pilot-scale contactor. The disinfection process was scale-able. An integrated ozone disinfection process consisting of a static mixer, gas-liquid separator and a plug flow contactor can achieve microbial inactivation that is comparable to an idealized well-mixed batch reactor. No significant microbial inactivation was observed at the outlet of the static mixer or the gas-liquid separator. In challenge experiments with *Bacillus subtilis* spores, *Cryptosporidium parvum* oocysts and *Giardia muris* cysts, there was little evidence of significant microorganism inactivation due to direct gas-liquid ozone contact within the static mixer itself. Although no microbial inactivation was measured within the static mixer itself, the static mixer provided the necessary starting conditions for efficient inactivation in the reactive flow segment and in the overall contacting system. These were (i) rapid dissolution of gaseous ozone and (ii) generation of a homogeneous dissolved ozone solution. The generation of a homogeneous dissolved ozone solution was possible due to the micromixing environment offered by the static mixer. This study also demonstrated that a static mixer, when integrated into a well-designed, continuous-plug flow ozone contacting system, can achieve excellent and predictable *Cryptosporidium parvum* inactivation. In such a system, *Cryptosporidium parvum* inactivation can be predicted using relatively straightforward modeling techniques.

The results of the laboratory study suggest that the potential role of the static mixer in improving microorganism inactivation was to serve as a means for rapidly and efficiently dissolving the gaseous ozone. Once the ozone was dissolved and the off-gas separated, efficient microorganism inactivation was achieved by contact with dissolved ozone in a well-designed reactive flow segment with an appropriate residence time distribution. When integrated into a two-stage contactor design, static mixers can, therefore, be effectively exploited to achieve efficient microorganism inactivation. This concept was supported conclusively by results of the testing with both *Bacillus subtilis* spores and *Cryptosporidium parvum* oocysts, and to a lesser certain extent with *Giardia muris* cysts, in the small-scale flow system used in the laboratory study. Although no meaningful enhancement was observed, inactivation of these microorganisms was equivalent to that predicted for contactor with perfect plug flow hydrodynamics. The implications of these findings are that the design criteria for optimum ozone dissolution are largely independent of the design criteria for optimum microbe reduction in a static mixer system.

The kinetics of inactivation of the cultured *Bacillus subtilis* spores (ATCC 6633) used in the laboratory and pilot-scale studies were characterized in batch reactor experiments. A model that described inactivation of the spores at temperatures ranging from 3°C to 22°C and pH values ranging from 6 to 8 was developed. Spore inactivation was found to be a non-linear function of the dissolved ozone Ct product with van't Hoff-Arrhenius type temperature dependence. Within the range of conditions investigated, inactivation was independent of pH and ozone concentration.

The modeling effort showed that Ct requirements for 1 log-unit inactivation of *Cryptosporidium parvum* or *Bacillus subtilis* are similar at all temperatures. A 2-log inactivation of *Cryptosporidium parvum* can be achieved at 2 to 2.3 times the equivalent *Bacillus subtilis* Ct value. Clearly, more experimental inactivation data is needed which should assist us to confirm the use of *Bacillus subtilis* as a suitable surrogate for *Cryptosporidium parvum,* especially at low temperatures. *Bacillus subtilis* spores are also useful for evaluation of the hydrodynamic efficiency of alternative ozone contactor designs, as they were found to be relatively resistant to ozone over a large temperature and pH range. The spores were used effectively to determine the efficiency of the laboratory-scale static mixer contactor and to investigate the potential effect of static mixer hydrodynamics on microorganism inactivation. Conclusions regarding the static mixer contactor that were arrived at using the spores were later confirmed with *Cryptosporidium parvum* oocysts and *Giardia muris*.

From an engineering design perspective, the advantage of the static mixer ozone contactor was that the mass transfer process was decoupled from the microorganism inactivation process. In the laboratory study static mixer contactor, ozone transfer efficiency was demonstrated to be essentially independent of microorganism inactivation. That is, the same level of inactivation was achieved with the same dissolved ozone concentration, but with either high or low mass transfer efficiency. In conventional fine bubble-diffusor contactors, gas-liquid mass transfer and microorganism inactivation are combined in the same vessels, making it difficult to optimize both simultaneously. In a static mixer contactor, on the other hand, the two processes may be physically separated and can, therefore, be optimized independently. In a full-scale contactor design, hydrodynamic conditions within the static mixer and the gas/liquid separator should be specified based on mass transfer requirements and related economics of ozone generation and use. The design of the reactive flow segment should be dictated by the requirements for microorganism inactivation. The design parameter that links the two processes is the dissolved ozone concentration. Either way, the results suggest that for a given ozone concentration, the most efficient operation is at a low *G/L* and high feed gas ozone concentration. This feature makes static mixer systems well-suited to the high concentrations and low gas flow associated with medium frequency ozone generators.

CHAPTER 3
PILOT-SCALE STUDY

OVERVIEW AND OBJECTIVES

The objective of the pilot study was to quantify the potential benefits of ozone application through the use of static mixers in terms of increased transfer efficiency, disinfection performance, and enhanced chemical reactions. Specific goals were to identify and quantify the effect of several water quality parameters (e.g., pH, DOC, alkalinity, temperature, etc) and environmental/engineered factors (e.g. ozone dose, and liquid velocity) on the disinfection capacity of the ozone-static mixer system.

These goals were achieved by conducting pilot studies with two different pilot units. The first one was built and implemented at the Neuilly-sur-Marne water treatment plant in France. The second pilot unit was built in collaboration with the University of Alberta as noted in the previous Chapter and was then operated at six different sites: Gwinnett County, GA (GC), Cleveland Department of Water, OH (CDW), Los Angeles Department of Water and Power, CA (LADWP), Chertsey Water Treatment Plant, UK (CWTP), Manatee Water Treatment Plant, FL (MWTP) and Elizabethtown Water Company, NJ (EWC). These sites allowed the testing of the ozone-static mixer system over a wide variety of operating conditions and source waters. The static mixer pilot unit was compared side-by-side to a pilot unit incorporating fine bubble diffusion contactors at LADWP and EWC to evaluate ozone dissolution performances, microbial inactivation, and bromate formation. Pilot tests allowed side-by-side process performance comparisons of static mixer systems and bubble contactor systems, as well as provided data for modeling reactor hydrodynamics and process performance of static mixers. *Bacillus subtilis* spores (as a surrogate for *Cryptosporidium*) were used for evaluating the effectiveness of the disinfection process. Pesticide destruction and bromate formation were studied under optimal conditions of microbial inactivation.

MATERIALS AND METHODS

The materials and methods used for the measurements of ozone and culturing/enumeration of *Bacillus subtilis* were the same as described in the previous Chapter. Water samples were shipped to University of Colorado Boulder for bromide/bromate measurements using ion chromatography. Potassium bromide was injected into raw water samples that did not contain detectable amounts of bromide.

FRENCH PILOT STUDY

Pilot Unit Description

The Générale-Des-Eaux pilot unit (5 to 20 m^3/h, 22 to 88 gpm, 6.8 m^3/h or 30 gpm typical) consisted of a skid which includes an oxygen fed ozone generator (2.2 kg/day or 4.8 lb/day at 10% O_3; Sorbios equipment), a static mixer (Sulzer) and a corresponding 4 m (13 feet) high gas/liquid separator column with no contactors. The skid also included analyzing equipment for analyzing gas and liquid ozone concentrations. The pilot unit was sized so that the results

Table 3.1
Characteristics of the Sulzer static mixer

Parameter	Value (Customary Unit)	Value (SI Unit)
Number of mixing elements	4	4
Length of one mixing element	2 inches	50 mm
Diameter of one mixing element	2.1 inches	54.5 mm
Number of voids	2	2
Length of voids	3.9 inches	100 mm
Total static mixer length	15.7 inches	400 mm
Length / Diameter ratio	7.34	7.34

could be extrapolated to a full-scale system. The Sulzer mixer was fitted with SMV-8 mixing elements of DN 50 (2 inches, 50 mm nominal diameter). Table 3.1 presents the main characteristics of the static mixer.

Assuming intermediate ozonation (ozonation prior to GAC and after sand filtration), the pilot unit was fed by clarified water from sand filtration. A significant bacterial population was present in the clarified water. This assisted us to better understand the impact of static mixers on disinfection. The water quality of the settled water is shown in Table 3.2. Figure 3.1 is a schematic of the pilot unit. The static mixer (horizontal) was located next to a gas/liquid separator of 250 mm (9.8 inches) diameter. The water level within the gas/liquid separator column could be adjusted through three draining-off heights (1.5 m (5ft), 2.5 m (8.2 ft) and 3 m (11.5 ft)).

The contact time varied from 30 to 100 seconds in the column and from 0.2 to 0.6 second in the static mixer depending on the hydrodynamic conditions. Ozone from oxygen was produced by a Sorbios generator delivering gas ozone concentration from 50 to 300 g O_3/m^3 NTP. Ozone was injected upstream of the static mixer. A back-flow preventer prevented any water getting into the ozone generator. The feed-gas and off-gas ozone concentration was measured by an UV/ozone analyzer. The ozone residual was measured at the static mixer outlet and at the gas/liquid separator column outlet, either by an Orbisphere analyzer or through the indigo method. The off-gas from the gas/liquid separator was destroyed through a thermal destruct unit. The pressure drop across the static mixer was measured as a function of the liquid flow rate through a mercury manometer.

Optimization of the Transfer Efficiency

Transfer Efficiency Statistical Approach

The influence of the following parameters on transfer efficiency was studied:

1. the liquid velocity (X_1, in m/s) through the static mixer or indirectly the pressure drop generated by the mixing elements;
2. the gas/liquid (X_2, flow rate ratio in %) ratio or indirectly the feed gas ozone concentration; and
3. the ozone residual (X_3, in mg/L) in the static mixer outflow. This variable has been preferred to the ozone dose, which is dependent on the ozone demand.

Table 3.2
Settled water quality parameters

Parameter	Winter	Summer
pH	7.5	7.5
Temperature (°C)	9.3	22.0
TOC (mg/L)	2.0	1.9-2.2
Turbidity (NTU)	0.5	0.31-0.35
UV_{254} (cm^{-1})	0.040	0.037-0.041
Alkalinity (mg $CaCO_3$/L)	200	175-185

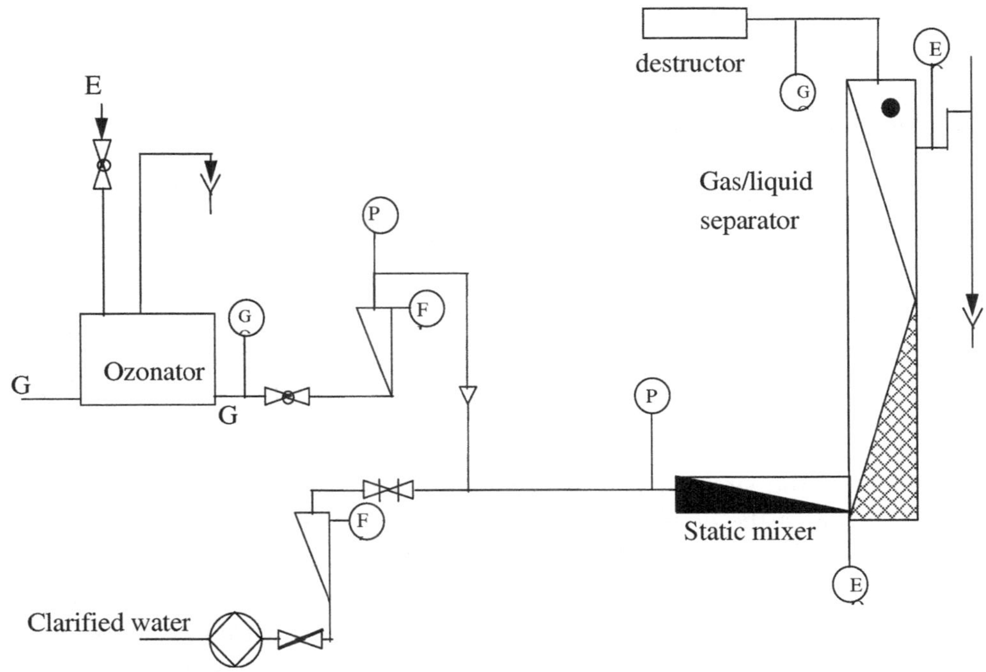

Figure 3.1 Schematic of the French pilot unit

The off gas ozone concentration was measured at the outlet of the gas/liquid separator which enabled us to determine the ozone transfer efficiency. The ranges for each variable are shown in Table 3.3.

Experimental Plan

Due to a large number of operating parameters, the utilization of Doehlert's experimental plan allowed to minimize the number of experiments leading to a mathematical model for each performance criterion. The Doehlert's method enables to link the performance criteria (transfer efficiency, disinfection efficiency, and bromate formation) and the operating parameters (ozone dose or ozone residual, liquid velocity, G/L ratio) through a mathematical expression. This technique allows in determining a second-degree mathematical function linking the operating

Table 3.3
Ranges of the operating variables for transfer efficiency testing

	Variable	Unit	Lower value	Upper value	Central value	Variation step
X_1	Liquid velocity	m/s	0.45	1.00	0.72	0.27
X_2	G/L Ratio	%	0.50	5.00	2.75	2.25
X_3	SM ozone residual	mg/L	0.10	2.00	1.05	0.95

variables, expressed in reduced coordinates (X_i), to the answer (Y). For example, with three variables the reduced coordinates are X_1, X_2, X_3, the answer Y is given by the equation:

$$Y = b_0 + b_1.X_1 + b_2.X_2 + b_3.X_3 + b_{11}.(X_1)^2 + b_{22}.(X_2)^2 + b_{33}.(X_3)^2 + b_{12}.X_1.X_2 + b_{13}.X_1.X_3 + b_{23}.X_2.X_3 \quad (3.1)$$

Transfer Efficiency Study

Following equation was obtained from the French study linking the transfer efficiency to liquid velocity (X_1), G/L ratio (X_2), and ozone residual (X_3).

$$\begin{pmatrix} \text{transfer} \\ \text{efficiency \%} \\ \end{pmatrix} = 81.8 + 11.0X_1 - 8.7X_2 - 8.5X_1^2 + 3.5X_2^2 + 1.4X_1X_2 + 1.5X_1X_3 + 2.0X_2X_3 \quad (3.2)$$

The analysis of the simplified model equation showed that the ozone residual (variable X_3) has little influence as the coefficients linked to this variable have been eliminated. On the other hand, the X_2 value (G/L ratio), will have to be minimized to get the higher transfer efficiency. The impact of the liquid velocity (X_1) on transfer efficiency is more difficult to estimate. Indeed, within the studied ranges, the increase in the X_1 variable can be either favorable or unfavorable to transfer: the coefficients b_1, b_{11}, b_{12} and b_{13} have some antagonistic effects.

It was possible to study accurately the influence of liquid velocity and G/L ratio on transfer efficiency by fixing the ozone residual (1 mg/L).

According to Figure 3.2, to reach an optimum transfer efficiency (mixer + degassing column), the G/L ratio must be as low as possible. An increase in liquid velocity is favorable to a transfer efficiency increase, nevertheless a plateau at 0.8 m/s is reached. Beyond this velocity and for the G/L ratios less than 3%, there is not much improvement in transfer efficiency. The goal is to find the maximum transfer efficiency while minimizing the pressure drop (see Figure 3.3). For a 1 mg /L ozone residual, the transfer efficiency reaches 95% when the G/L ratio is less than 1% and the liquid velocity is greater than 0.75 m/s (pressure drop=100 mbars, headloss = 1 m or 3.4 feet). To reach transfer efficiencies greater than 90%, the maximal G/L ratio is 1.5% and the required liquid velocity is about 0.7 m/s (pressure drop=80 mbars, headloss = 2.7 feet).

While using a static mixer configuration of 4 mixing elements, the optimal transfer efficiency is reached with a low G/L ratio. An increase in liquid velocity is also favorable to optimize this parameter. Nevertheless, a plateau at 0.8 m/s is noticed with a G/L ratio less than 3%. Beyond this velocity, there is no improvement of transfer efficiency. This velocity

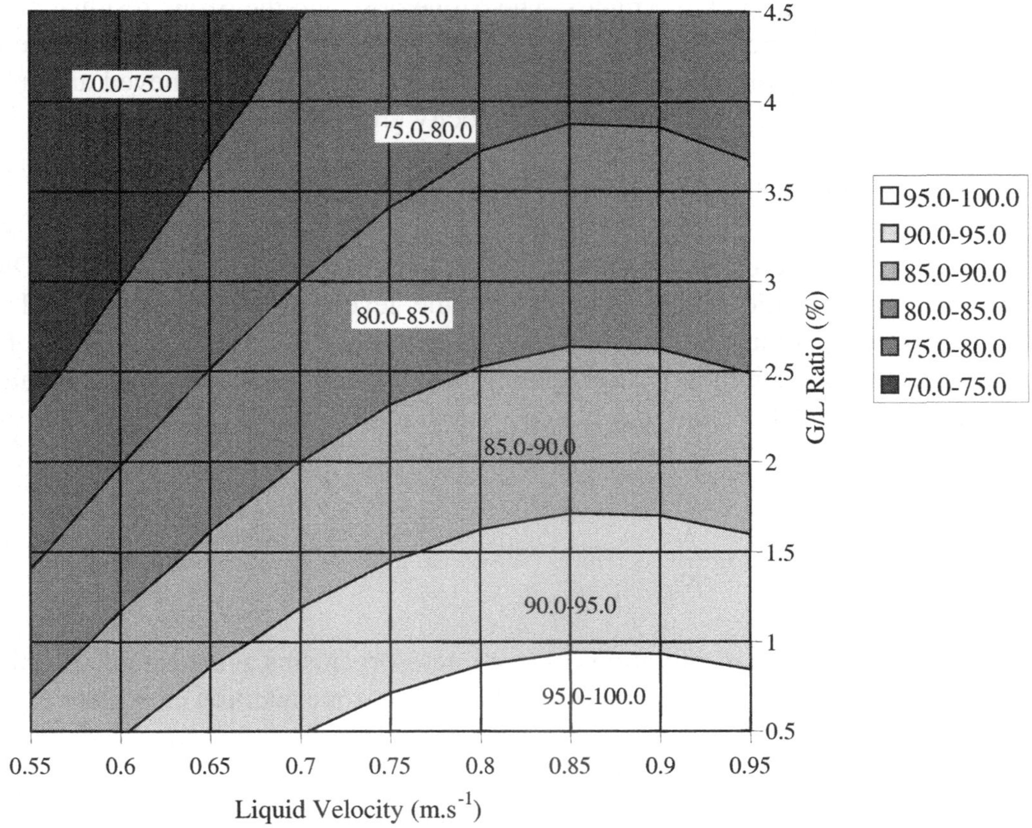

Figure 3.2 Influence of the liquid velocity and G/L ratio on the transfer efficiency (ozone residual at the static mixer outlet = 1 mg/L)

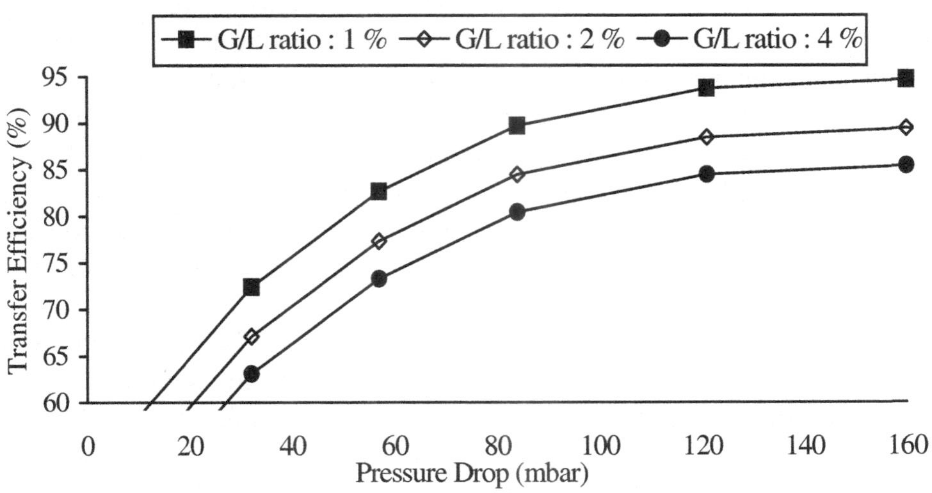

Figure 3.3 Influence of G/L ratio and pressure drop on transfer efficiency (ozone residual at the SM outlet = 1 mg/L)

corresponds to a pressure drop of 100 mbars. The ozone dose and the ozone residual at the outlet of the static mixer have little influence on the transfer efficiency compared with the first two variables. With this configuration, the transfer efficiencies that can be achieved are greater than 90% with a G/L ratio close to 1.5% and a liquid velocity greater than 0.75 m/s.

Disinfection Performance

Disinfection tests were first conducted on HPC bacteria at warm temperature. Other trials were then performed at cold temperature, where the levels of HPC bacteria and indicator microorganisms of fecal contamination were measured both at concentrations typical of clarified water and at increased concentrations after spiking the clarified water with sewage water. Along the process, spore inactivation was calculated as log inactivation according to the following equation:

$$\text{log inactivation} = -log\frac{N}{N_0} \tag{3.3}$$

N and N_0 were the measured concentrations of live micro-organisms at the sample location and in the feed water (after micro-organisms addition but prior to ozone addition), respectively.

Inactivation of HPC at a Temperature of 22°C

There are many parameters that can influence disinfection for a given static mixer. Such parameters include the headloss through the static mixer (or liquid velocity), the gas/liquid ratio and the dissolved ozone residual at the outlet of the static mixer. These parameters play an important role in determining the total transfer yield (Langlais et al., 1998), and have an interesting influence on disinfection efficiency. The range of the parameters tested is displayed in Table 3.4.

Velocities of 0.7 to 1.5 m/s correspond to respective headlosses of 127 to 592 mbars. Tests were performed at a temperature of 22°C. From the results of the tests, the model obtained to predict the HPC inactivation had a standard deviation of ± 0.24 log-units. The model incorporated the effects of liquid velocity and the ozone residual to predict the HPC inactivation.

Then the impact of the gas/liquid ratio was analyzed. Figure 3.4 showed that an increase in liquid velocity had a negligible impact on disinfection, once the ozone residual is greater than 0.8 mg/L. At low ozone residuals (lower than 0.5 mg/L), the disinfection is improved as the liquid velocity decreases. This may be due to higher contact times, which enabled extended inactivation of microorganisms.

The impact of the gas/liquid ratio is shown graphically in Figure 3.5 and Figure 3.6. Here the results were obtained using different dissolved ozone residuals at the outlet of the static mixer.

For a dissolved ozone residual of 0.9 mg/L, the best disinfection conditions are created when there is a small gas/liquid ratio and when liquid velocity is low. With a dissolved ozone residual of 0.2 mg/L however, the tendency is reversed. Increasing the gas/liquid ratio and the liquid velocity improves HPC inactivation, although not quite to levels observed when ozone residual is 0.9 mg/L.

The laboratory results above are only valid for temperatures of 22 °C. At 22 °C, with oxygen-produced ozone and a static-mixer followed by a contact chamber, HPC inactivation can

Table 3.4
Range of the parameters for HPC inactivation at a temperature of 22°C

Parameter	Minimum	Average	Maximum
Liquid velocity (m/s)	0.70	1.10	1.50
Gas/liquid velocity (%)	1.75	3.62	5.50
Dissolved ozone residual (mg/L)	0.10	0.55	1.00

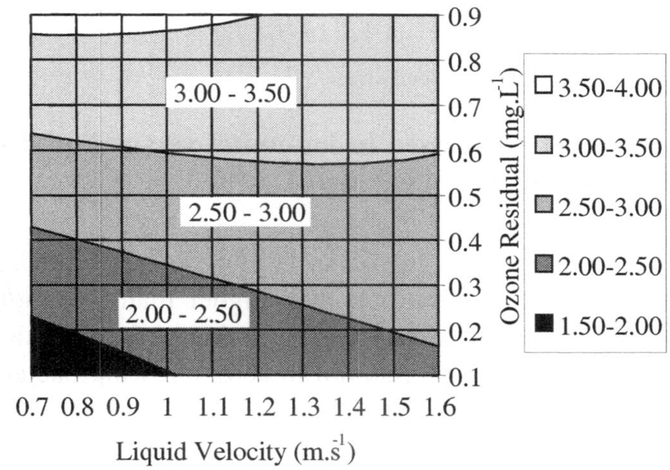

Figure 3.4 HPC inactivation (expressed in log-units) versus liquid velocity and dissolved ozone residual for gas/liquid ratio of 2%

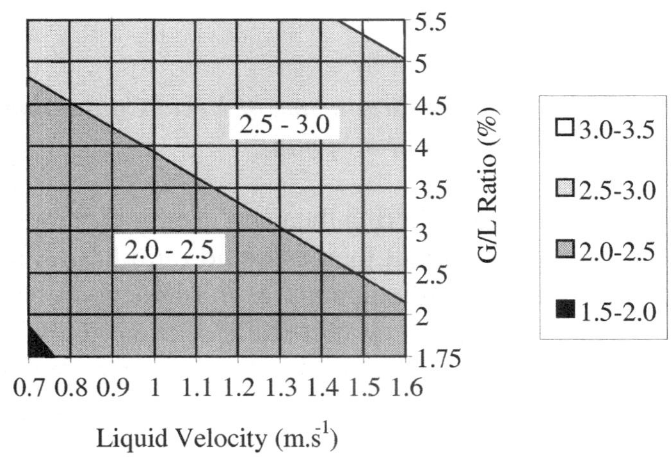

Figure 3.5 HPC inactivation (expressed in log-units) versus liquid velocity and gas/liquid ratio, for a dissolved ozone residual of 0.2 mg/L

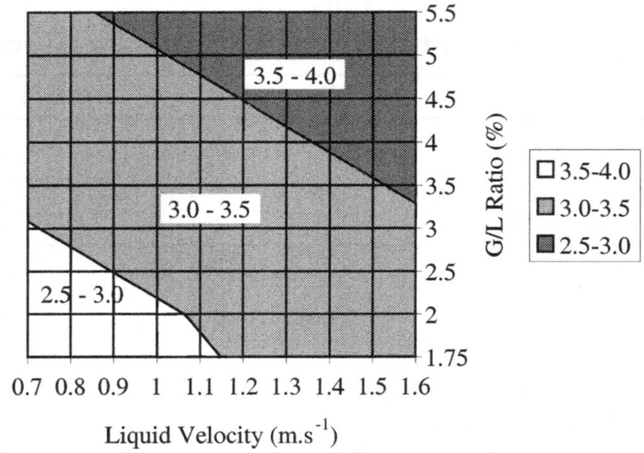

Figure 3.6 HPC inactivation (expressed in log-units) versus liquid velocity and gas/liquid ratio, for a dissolved ozone residual of 0.9 mg/L

reach 3 log-units. This assumes an ozone residual higher than 0.6 mg/L and a gas/liquid ratio lower than 3 %. The ozone residual cannot be lower than 0.6 mg/L, or else the residual at the outlet in the contact chamber may, in places, fall to zero, allowing bacteria regrowth. The results also indicated that lower gas/liquid ratios increase the transfer yield, as does using oxygen rather than air to produce ozone. At colder temperatures disinfection is more difficult, although it is interesting to observe the changes in HPC inactivation compared with results obtained at warm temperature.

Table 3.5 shows the results obtained with two values of ozone residual at the outlet of the static mixer (0.35 and 0.80 mg /L) on settled water.

The disinfection of these microorganisms was effectively absolute and thus the influence of the ozone residual could not be described. Sewage water containing microorganisms was therefore injected into the sample.

Inactivation of HPC and Indicator Microorganisms of Fecal Contamination at a Temperature of 9 °C (With Sewage Water Spiking)

Spiking was carried out without altering the main chemical parameters of the water (TOC, pH, Turbidity and UV 254 nm). Given that the gas/liquid ratio has a limited influence on HPC bacteria when using static mixer at warm temperatures in such conditions (ozone produced from oxygen), the influence of that parameter was not studied, and was instead fixed at 2 %. Only liquid velocity and ozone residual at the outlet of the static mixer were investigated. The range of the two parameters is shown in Table 3.6. A different model was obtained for each microorganism measured. As the trends are quite similar from one type of microorganism to another, only some results (fecal coliforms, fecal streptococci and HPC) are presented. The standard deviation calculated for each model is summarised in Table 3.7.

The different figures (see Figures 3.7, 3.8, and 3.9) show that ozone residual is the most influential parameter in improving disinfection in static mixers. A decrease of liquid velocity from 1.9 to 0.9 m/s also improves the inactivation of these HPC bacteria and fecal coliforms.

Table 3.5
Inactivation of HPC and indicator microorganisms of fecal contamination at a temperature of 9°C (without spiking)

Type of micro-organism	Settled water	Ozone residual 0.35 mg/L	Ozone residual 0.80 mg/L
Total coliforms (#/100mL)	1375	12	3
Fecal coliforms (#/100mL)	380	1	2
Fecal streptococci (#/100mL)	20	0	0
Total germs 24 h (#/mL)	62	2	2
Total germs 72 h (#/mL)	364	5	4
HPC bacteria (#/mL)	35450	20	26

Note: Micro-organisms concentration at the outlet of the static mixer, gas/liquid ratio of 1.60 %, liquid velocity of 0.77 m/s.

Table 3.6
Range of the parameters for disinfection efficiency study at a temperature of 9°C

Parameter	Minimum	Average	Maximum
Liquid velocity (m/s)	0.70	1.35	2.00
Dissolved ozone residual (mg/L)	0.10	0.55	1.00

Table 3.7
Standard deviation according to microorganism type

	Fecal coliforms	Fecal streptococci	HPC bacteria
Standard deviation	± 0.10 log-units	± 0.15 log-units	± 0.10 log-units

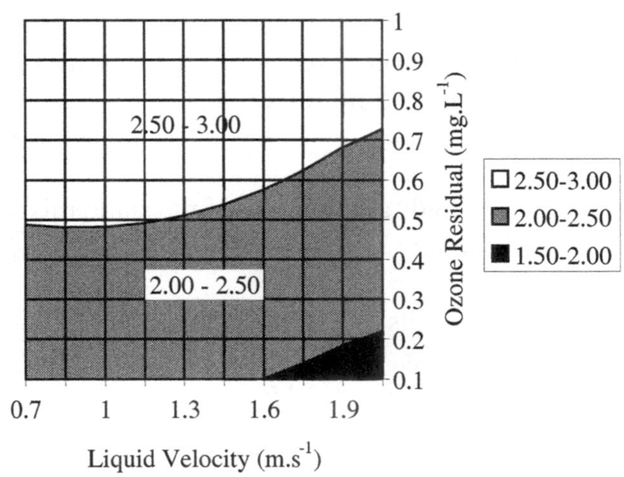

Figure 3.7 Fecal coliform inactivation (expressed in log-units) versus the liquid velocity and the residual ozone at the outlet of static mixer

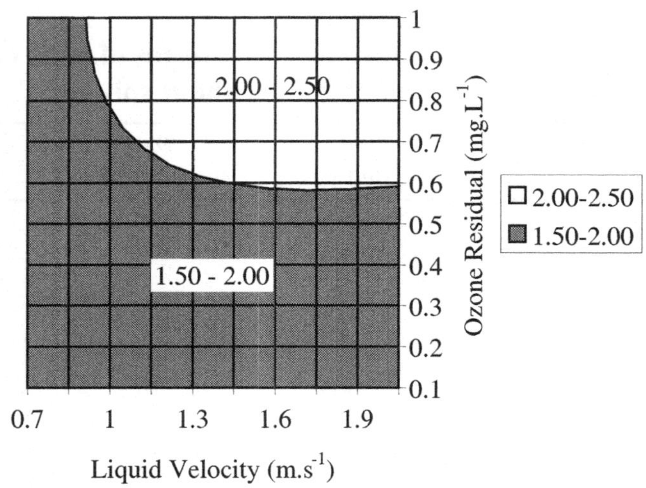

Figure 3.8 Fecal streptococci inactivation (expressed in log-units) versus the liquid velocity and the residual ozone at the outlet of static mixer

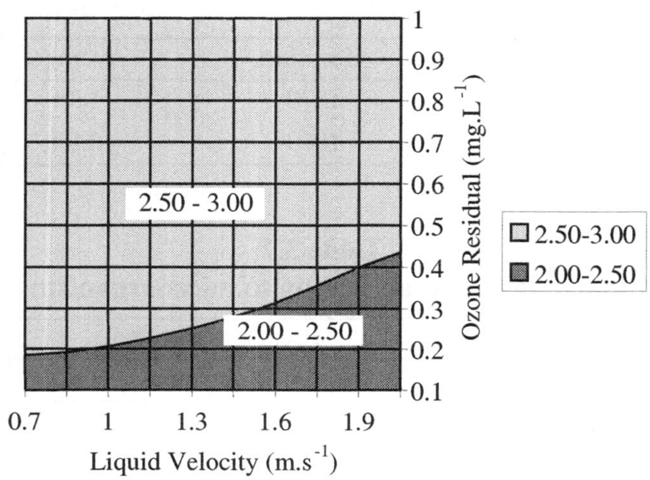

Figure 3.9 Inactivation of HPC bacteria (expressed in log-units) versus the liquid velocity and the residual ozone at the outlet of static mixer

With an ozone residual of 1 mg/L at the outlet of the static mixer, the best results obtained for the following types of bacteria were:

1. Fecal coliforms: 3.00 log-units ± 0.10
2. Total coliforms: 2.70 log-units ± 0.10
3. Fecal streptococci: 2.20 log-units ± 0.15
4. HPC (24 h and 37°C): 2.20 log-units ± 0.25
5. HPC (72 h and 24°C): 2.20 log-units ± 0.25

Further inactivation in bacteria could be expected if the ozone residual at the outlet of the mixer increases. Changes in ozone residual and liquid velocity have comparable effects on HPC inactivation at different temperatures (i.e. 9°C or 22°C). However, the inactivation of bacteria is more efficient at higher temperatures. For example, the difference in inactivation between ozonation with an ozone residual of 0.8 mg/L performed at 22°C and at 9°C is around 0.4 log-units.

Atrazine Oxidation

Methods

Settled water was spiked with atrazine to reach a concentration of 800 ng/L. Assays were performed with a water quality of settled water as presented in Table 2.2 during summer.

Hydrogen peroxide (H_2O_2) was injected at each run at a constant H_2O_2/O_3 ratio equal to 0.4 g/g. To determine and optimize the oxidation efficiency of the static mixer unit, the main operating parameters were the following:

1. Liquid velocity across the static mixer between 0.7 m/s and 1.5 m/s
2. G/L ratio between 1.75% and 5.50% (this parameter is directly linked to the feed-gas ozone concentration),
3. Applied ozone dose between 1.5 mg/L and 5.0 mg/L

Samples were taken at the outlet of the static mixer and at the outlet of the separator column to quantify the atrazine oxidation. A statistical approach was followed in order to characterize the influence of each operating parameter on oxidation effectiveness.

Atrazine Reduction Through the Static Mixer

Figure 3.10 proves that atrazine oxidation within the static mixer is strongly dependent on the applied ozone dose. With an applied ozone dose of 5 mg/L and a H_2O_2/O_3 ratio equal to 0.4 g/g, the oxidation rate reaches 60% to 70%. The feed-gas ozone concentration (or the G/L ratio) has little influence on this oxidation efficiency when using an applied ozone dose higher than 3.5 mg/L.

Regarding the liquid velocity (or the pressure drop) through the static mixer, Figure 3.11 proves this operating parameter has a limited influence on atrazine oxidation.

Atrazine Reduction Through the Overall Dissolution System (Static Mixer + Separator Column)

Figure 3.12 confirms that applied ozone dose is the dominant operating parameter for atrazine oxidation. Nevertheless a higher liquid velocity across the static mixer slightly favors atrazine oxidation since the transfer efficiency and the mass transfer coefficient k_La are increased. This increase in oxidation efficiency is observed in spite of the decrease in contact time within the dissolution system (Table 3.8).

Atrazine oxidation being a fast-kinetic reaction, the gas-liquid contact conditions are crucial to reach a high oxidation effectiveness. As a static mixer provides a high mass transfer coefficient k_La, its effectiveness for oxidizing pesticides is expected to be better than a conventional ozone contactor fitted with porous diffusers.

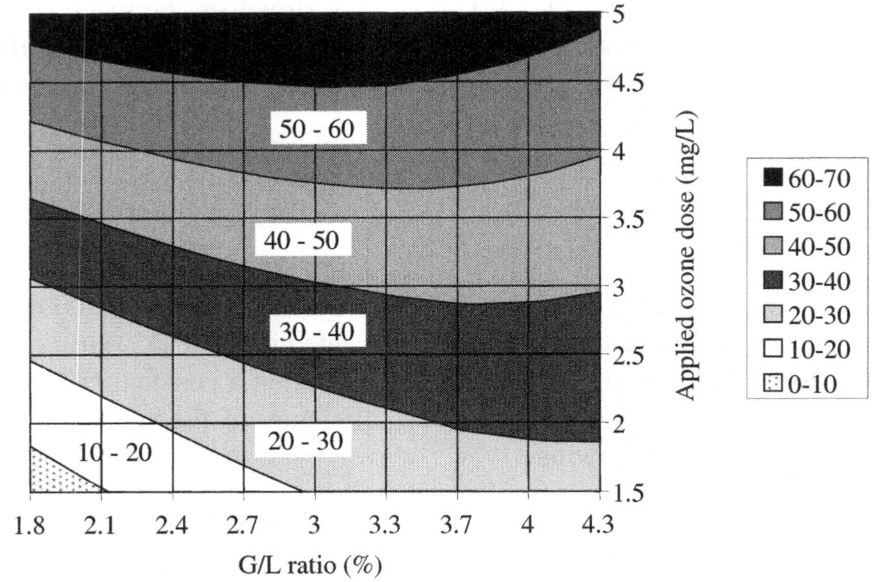

Figure 3.10 Combined influence of applied ozone dose and G/L ratio on atrazine oxidation at the outlet of the static mixer

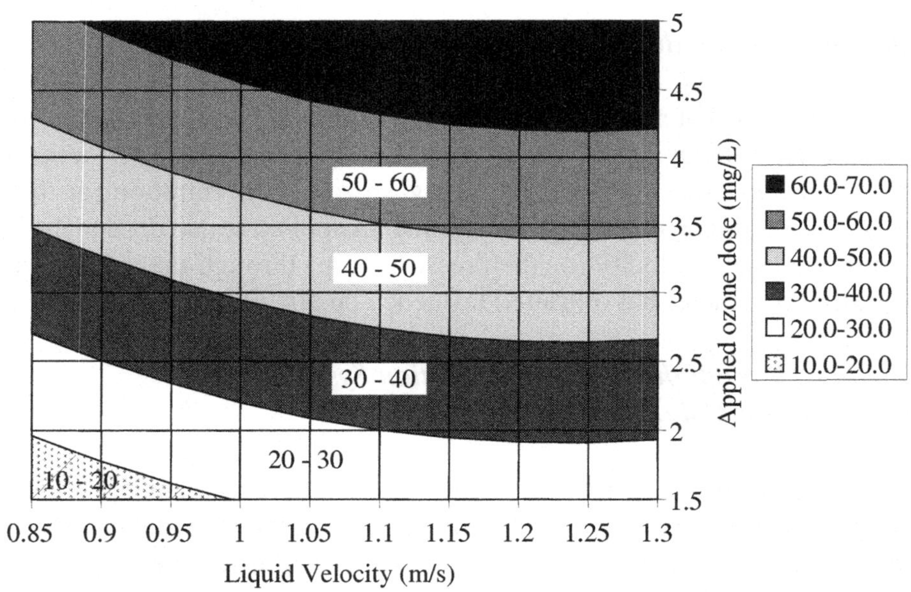

Figure 3.11 Combined influence of applied ozone dose and liquid velocity on atrazine oxidation through the outlet of the static mixer

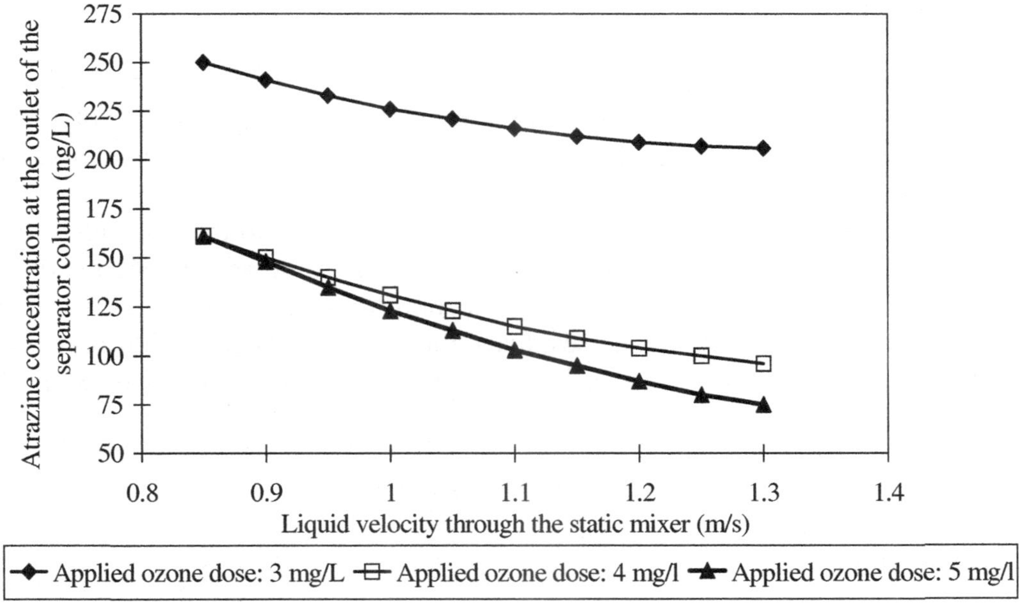

Figure 3.12 Atrazine residual at the outlet of the separator column with an initial atrazine concentration of 800 ng/L and a G/L ratio of 2%

Table 3.8
Hydraulic Residence Time (HRT) within the separator column according to the liquid velocity through the static mixer

Liquid velocity (m/s)	Pressure drop (psi / KPa)	HRT within the separator column (s)
0.85	1.8 / 12.5	62
0.90	2.0 / 14.1	58
0.95	2.3 / 15.7	55
1.00	2.5 / 17.4	52
1.05	2.8 / 19.2	50
1.10	3.1 / 21.1	48
1.15	3.4 / 23.1	45
1.20	3.6 / 25.1	44
1.25	3.9 / 27.2	42
1.30	4.3 / 29.4	40

Bromate Formation

The bromate formation inside the static mixer was investigated with a bromide concentration of 40 µg/L in settled water (temperature = 22°C). An experimental plan was created with gas/liquid ratio, liquid velocity and ozone residual at the outlet of the static mixer, being the three chief variables (Table 3.9).

Variations in the gas/liquid ratio corresponded to variations in the concentration of ozone gas from 22 to 175 mg O_3/L. The raw water quality was the same as described earlier. In most cases the bromate concentration was below the detection limit (2 µg/L) and even in particularly bad conditions (low velocity, high ozone residual), the bromate concentration remained around 3 µg/L. A decrease in gas/liquid ratio (i.e. the ozone gas concentration) caused no observed increase in bromate formation.

UNITED STATES AND UNITED KINGDOM PILOT STUDY

Pilot Unit Description

The ozone contacting system used for the US and UK pilot studies consisted of a 15.7 mm ID static mixer for rapid ozone dissolution, a vertical gas/liquid separator (ID = 102 mm) for additional ozone dissolution and gas-liquid separation, and four contact columns (ID = 203 cm) to provide extended contact time between dissolved ozone and microbes to be inactivated (Figure 3.13). A photograph of the unit is shown in Figure 3.14. As noted in the previous Chapter this unit was built along with the laboratory unit at the University of Alberta. Ozonated gas, produced by an oxygen-fed ozone generator (Praxair-Trailigaz), was introduced into the center of the 1.3 mm ID pipe directly upstream of the static mixer. Side-by-side testing between the Static Mixer Unit (SMU) and a Fine Bubble Diffuser Unit (FBDU) was performed at LADWP and EWC. Main operational parameters are given in Table 3.10. These parameters were fully characterized and optimized in order to:

1. achieve a ozone transfer efficiency of greater than 90%, while minimizing the pressure drop across the static mixer. Langlais et al. (1998) noted that low gas to liquid flowrate (G/L) ratios result in high transfer efficiencies. Hence, the highest possible feed gas ozone concentration was applied (6% wt.) to lower the gas flowrate to be injected for applying a given ozone dose.
2. ensure a hydrodynamic behavior close to a plug flow behavior: tracer testing showed a t_{10} / t ratio = 0.87 (next section) for the overall contact columns of the SMU.

The general arrangement of the FBDU (3 contactors total) at EWC was similar to the one at LADWP. The flowrate was kept constant to 7.6 L/min (2 gpm). Each column provided a 7.2 minutes contact time at this liquid flowrate. Thus, the overall contact time through the FBD unit was equal to 22 minutes. The ozone generation system of this unit consisted of an oxygen concentrator feeding an ozone generator able to provide an ozone concentration of 18 g/Nm^3 (1.2 %wt.). The flowrate through the SMU was kept constant at 11.4 L/min (3 gpm). Liquid velocity across the static mixing elements was, therefore, equal to 0.97 m/s resulting in a 2 psi pressure drop across the static mixer. Main operating parameters used during the comparison tests are shown in Table 3.11.

Table 3.9
Range of parameters for bromate formation

Parameter	Minimum	Average	Maximum
Liquid velocity (m/s)	0.70	1.45	2.00
Gas/liquid ratio (%)	0.50	2.75	5.00
Dissolved ozone residual (mg/L)	0.10	0.55	1.00

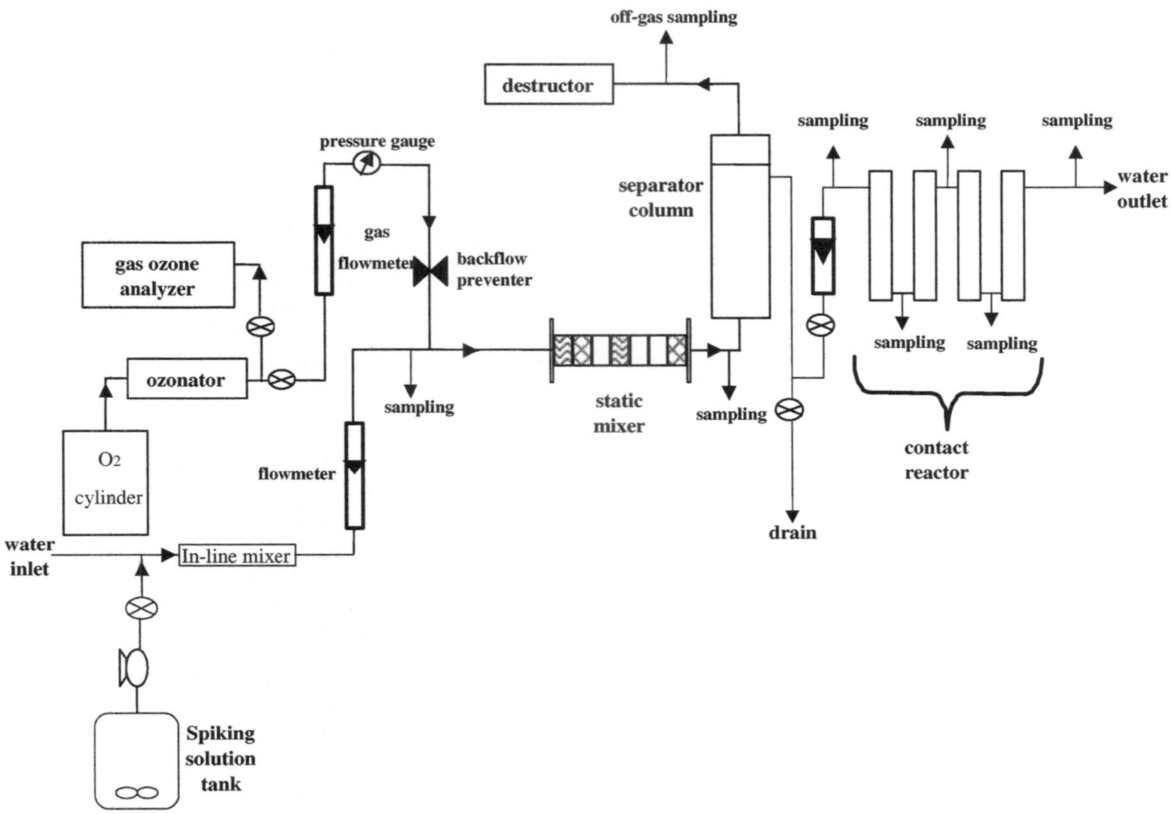

Figure 3.13 Schematic of the pilot-scale static mixer unit

Residence Time Distribution Analysis

Tracer tests were conducted on both the separator and contact columns in order to study the influence of contactor hydrodynamics on disinfection effectiveness. For the separator column tests, tracer was injected upstream of the static mixer. In this way, the RTD of the static mixer itself was incorporated into the separator column test. Because the hydraulic residence time of the static mixer was less than a second, it was not deemed feasible or worthwhile to determine the RTD of the static mixer separately. The downstream concentration of tracer was measured at a sample point close to the top of the liquid level in the separator column. Tracer tests were carried out at flow rates of 11.4 L/min (3 gpm) and 22.7 L/min (6 gpm). Air was injected in place of ozonized gas and the G/L ratio was 3%. Each test was duplicated. Figure 3.15 presents the RTD

Figure 3.14 Photograph of the pilot-scale static mixer unit

curves and Table 3.12 shows the main hydrodynamic parameters calculated from these RDT curves. The calculation of the liquid volume accounted for the volume of the piping between the tracer injection point and the inlet of the separator column. The contact time of this piping including the static mixer was respectively equal to 4 sec and 8 sec at 22.7 L/min (6 gpm) and 11.4 L/min (3 gpm).

For the contact columns the tracer was injected at the inlet to the first contact column and downstream samples were collected at the outlet of the first contact column, the outlet of the second contact column and the outlet of the fourth contact column. Tracer testing was carried out with a 11.4 L/min (3 gpm) flow rate. The test was duplicated. Figure 3.16 shows the RTD curves along the contactor. Table 3.13 lists the main hydrodynamic parameters calculated from the RTD curves. The reason for conducting independent tracer tests on the separator and contact columns was to be able to understand the hydrodynamics of the different components of the pilot static

Table 3.10
Operating parameters of the static mixer unit and the fine bubble diffuser unit (LADWP)

Parameter	Value
SMU – Gas Dissolution System*	
Liquid flowrate, G/L ratio (%)	15.1 L/min (4 gpm), 1.5 to 2.7
Pressure drop across the static mixer	3 psi
Contact time through the static mixer	0.2 s
Contact time through the gas-liquid separator	1.5 min
Ozone dose	0.6 mg/L to 2.0 mg/L
SMU – Contactor	
Liquid flowrate	11.4 L/min (3 gpm)
Ozone residual inflow	0.5 mg/L to 1.5 mg/L
Contact time per column	6 min
Overall contact time	25.6 min
FBDU – Gas Dissolution System*	
Liquid flowrate, G/L ratio (%)	7.6 L/min (2 gpm), 2.5-5.7
Contact time through the gas-liquid column	6.2 min
Ozone dose	1.0 mg/L to 1.7 mg/L
FBDU – Contactor	
Ozone residual inflow	0.5 mg/L to 1.5 mg/L
Contact time through the gas-liquid column	6.2 min
Overall contact time	19 min

* For the SMU, the dissolution system consists of the static mixer and the gas-liquid separator, whereas for the FBDU it consists of the first column fitted with porous diffusers.

Table 3.11
Operating parameters of both the SMU and the FBDU-EWC

	Static Mixer Unit	Fine Bubble Diffuser Unit
Liquid Flowrate (L/min)	11.4	7.6
Overall Contact Time (min)	26	22
Contact through the dissolution system (min)*	2	7.2
G/L ratio (%)	3.1 to 5.0	11.1 to 14.4

* For the SMU, the dissolution system consists of the static mixer and the gas-liquid separator, whereas it consists of the first column for the FBDU.

mixer system. The overall RTD of the system can be derived by combining the RTDs of the separator and contactor sections.

The residence time distributions show that the hydraulic regime in both columns was typical of plug flow with dispersion. All indications are that approximation to plug flow was achieved in the pilot-scale static mixer system. The measured average residence times, t_m, and the theoretical residence times, τ, are close and the T_{10}/τ ratios were greater than 0.80. In addition, there was

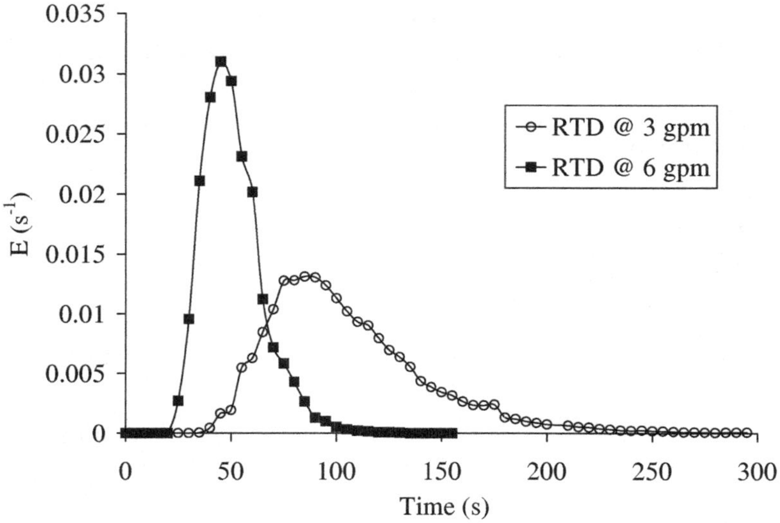

Figure 3.15 RTD curve of the separator column at 11.4 L/min (3 gpm) and 22.7 L/min (6 gpm) using methlyene blue tracer with air addition (G/L ratio = 3%)

Table 3.12
Hydrodynamic parameters of the separator column

Hydrodynamic Parameters	Q = 11.4 L/min (3 gpm)	Q = 22.7 L/min (6 gpm)
τ (s)*	88	44
T_{10} (s)	65	34
T_{90} (s)	157	70
t_m (s)	106	51
T_{10} / τ	0.74	0.77
σ^2, (s^2)	1440	207

* τ = Theoretical hydraulic residence time

Table 3.13
Hydrodynamic parameters of the contact columns

Hydrodynamic Parameters	1st column	1st + 2nd column	Overall contact reactor
τ (min)	6.0	12.0	24.0
T_{10} (min)	5.2	9.4	20.9
T_{90} (min)	8.0	15.1	29.5
t_m (min)	6.3	11.9	24.9
T_{10} / τ	0.87	0.78	0.87
σ^2 (s^2)	6454	18558	40556

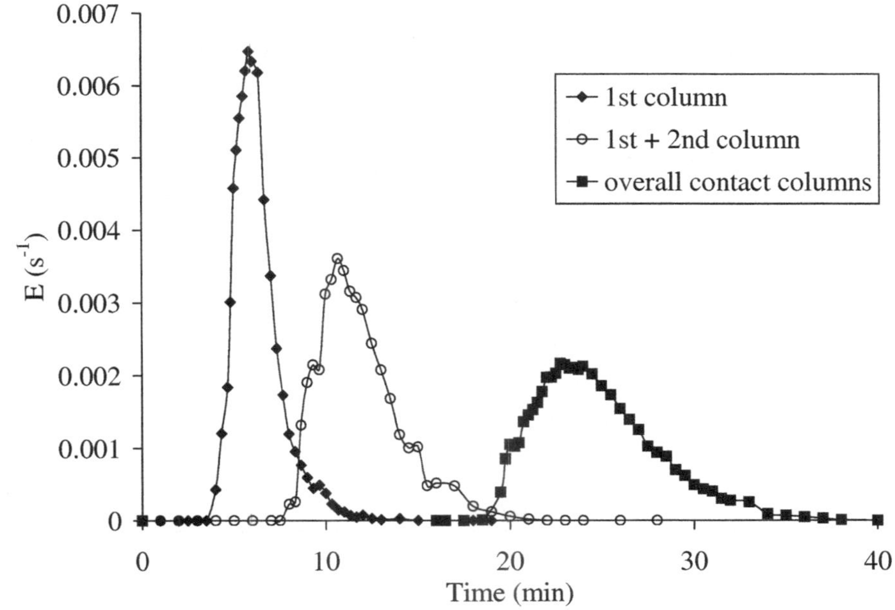

Figure 3.16 RTD curves along the contact columns at 11.4 L/min

no evidence of flow by-passing or short-circuting either in the measured RTD curves, or by direct visual observation of the colored tracer during the tests. There was evidence of greater back-mixing in the separator column. This was attributed to the two-phase flow condition. The residence time of the contact columns (about 25 min) was many times greater than that of the separator column (about 1 min), therefore, the RTD of the overall system was dominated by the contact columns. The RTD of the overall static mixer system was, therefore, close to plug flow with low dispersion. This plug flow characteristic was necessary to achieve a high level of microbial inactivation in the flow system. The measured RTDs suggest that the systems may be successfully modeled as plug flow with dispersion.

Water Quality Parameters for the Pilot Study

Water quality parameters for each pilot-study site are shown in Table 3.14. At the MWTP, the SMU was fed by the effluent from a coagulation/flocculation/sedimentation (Actiflo®) pilot unit.

Transfer Efficiency Performances and Ozone Profiling

Transfer Efficiency Study

Ozone concentration in the feed-gas ($C_{O3,f}$) and the off-gas ($C_{O3,o}$) was measured by the potassium iodide absorption method (Langlais et al. 1991). Liquid samples were collected along the pilot unit at various operating conditions for ozone residual measurement using the indigo trisulphonate colorimetric method. Vaccum ampules (Accuvac; Hach Company) were also used to measure the dissolved ozone residual concentration as a quality control on selected samples.

Table 3.14
Water quality parameters at the various pilot-study sites

Water-Quality Parameter	Gwinnett County Filtered Water	Tampa Bay Actiflo Effluent	Los Angeles Raw Water
DOC (mg/L)	1.4 to 3.2	3.1 to 4.1	1.7 to 2.0
UV_{254} (cm^{-1})	0.010 to 0.016	NA	0.052 to 0.053
Alkalinity (mg/L)	17 to 18	1 to 11.6	106 to 109
Turbidity (NTU)	0.2 to 0.6	0.80 to 1.25	1.6 to 2.0
Temperature (°C)	17 to 19	20 to 21	14
Bromide (µg/L)	ND	115	95
pH	6.6 to 7.0	5.5 to 6.5	8.1
Water-Quality Parameter	**City of Cleveland Raw Water**	**Elizabethtown Raw Water**	**Chertsey Filtered Water**
DOC (mg/L)	2.3 to 2.5	2.6 to 5.4	2.5 to 2.7
UV_{254} (cm^{-1})	0.043 to 0.145	0.076 to 0.115	0.133 to 0.140
Alkalinity (mg/L)	78 to 90	38 to 47	230 to 236
Turbidity (NTU)	5.95 to 48.3	6.4 to 14.0	0.10 to 0.13
Temperature (°C)	8.5 to 11.5	5.0 to 6.1	18
Bromide (µg/L)	ND	ND	125
pH	8.0 to 8.2	7.1 to 7.2	7.3

ND = non detectable

Transfer efficiency (*TE*) was calculated as follows:

$$TE(\%) = 100 \times \frac{C_{O3,f} - C_{O3,o}}{C_{O3,f}} \quad (3.4)$$

Preliminary testing showed that a flowrate of 15.1 L/min (4 gpm) passing through the dissolution system (static mixer + separator), and a 11.4 L/min (3 gpm) flowrate feeding the contactors (1 gpm was wasted) would allow us to minimize the pressure drop and achieve optimal transfer efficiencies. At that flowrate, the pressure drop across the static mixer was 3 psi (3 mixing elements). The flowrate of 11.4 L/min across the contactors provided an overall contact time of 25.6 minutes and a plug flow behavior determined through the tracer tests as described earlier. Three levels of ozone dose were tested at each site. Transfer efficiency results are shown in Figure 3.17. Transfer efficiency increased as the G/L ratio (ozone dose) decreased. For the waters tested, a G/L ratio of 0.6 to 1.5 % and a liquid velocity of 1.3 m/sec resulted in the highest transfer efficiency (96% to 99%).

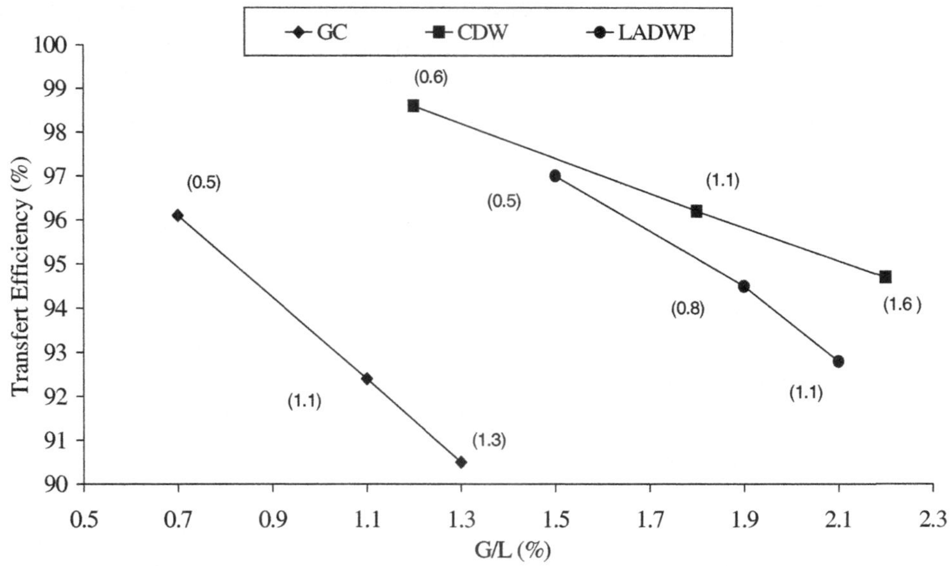

Figure 3.17 Transfer efficiency versus G/L ratio for different source waters. Numbers in parenthesis refer to ozone residual entering the contactors in mg/L.

Ozone Profiling

Ozone residuals along both the SMU and the FBDU allowed determination of the Cts of each ozone contacting system. Average Ct product at each sample location was calculated using the following equation:

$$Ct = \sum_{i=1}^{n} C_{avg,i} \times \left(HRT_i - HRT_{i-1}\right) \tag{3.5}$$

HRT_i : hydraulic residence time at the sampling location "i"
$C_{avg,i}$: average ozone residual between the sampling locations "i-1" and "i"

$$C_{avg,i} = \frac{\left(C_{avg,i-1} + C_{avg,i}\right)}{2} \tag{3.6}$$

Dissolved ozone concentrations along the contact columns obtained at each site have been used to study the ozone decay kinetics. In a perfectly mixed batch system, ozone decay follows a first-order kinetics:

$$C = C_0 \times \exp(-k_d t) \tag{3.7}$$

The pilot unit being a flow-through system, hydrodynamic behavior has to be taken into account to determine that first-order kinetic constant. Thus, the segregated flow analysis (Levenspiel, 1972) can be applied to link ozone decay kinetics and hydrodynamic behavior, which has been determined through tracer testing. Along an elementary segment, the reaction is assumed to take place in plug-flow and the concentrations vary as a function of the fluid residence time as in a batch reactor. At the outlet of the segment "residence time t", the concentration is C(t), as given by Equation 3.7. By putting together each segment, the concentration at the outlet of the flow-through system is:

$$C = \int_0^{\infty} C_0 \times \exp(-k_d t) E(t) dt \qquad (3.8)$$

$E(t)$: Residence Time Distribution
C : ozone residual at the outlet of the system
C_0 : ozone residual at the inlet of the system
k_d : ozone decay kinetic constant (s^{-1}).

Ozone residuals were measured at the outlets of the first, second and the fourth contact columns. A linear regression of the data resulted in a single ozone decay kinetic constant (k_d) for each trial (Table 3.15). For the water sources (Elizabethtown and Los Angeles), the ozone decay was significantly more rapid (higher kinetic constant). This observation can be explained by the chemical interactions of ozone with water quality parameters (e.g. TOC), and its reaction pathways (direct versus indirect) which are influenced by pH and radical scavengers such as alkalinity. Figure 3.18 shows the good agreement between predicted and experimentally obtained ozone residuals for the different waters tested.

Disinfection Performance

The pilot study evaluated trip controls, process controls, and three replicates at each ozone dose. Each microbial sample was analyzed in duplicate over two decimal dilutions. Spiking protocol was adjusted to get a concentration of 10^4 *Bacillus subtilis* spores/mL at the inlet of the process located upstream of the injection point of the ozonated gas. Prior to any disinfection trial, a high ozone dose (about 4 mg/L) was applied for 45 minutes to decontaminate the overall unit. At the end of each disinfection trial, the same decontamination procedure was followed. Water samples were collected at the various sampling points as shown in Figure 3.13 and analyzed for microorganisms. The results did not reveal any microbial contamination.

The inactivation data obtained with various source water qualities are shown in Figure 3.19. The temperature of the water tested played a significant role in microbial inactivation. By plotting the inactivation results as a function of Ct, water quality parameters are taken into account since ozone residual (C) decay is influenced by these parameters. Table 3.16 shows Ct values required to reach different levels of inactivation of *Bacillus subtilis* spores at the various sites tested. Comparing the results at Gwinnett and Cleveland, we see that when the temperature is reduced by half, the ozone dose (mg O_3/mg C) and Ct required to achieve similar level of microbial inactivation were almost doubled.

When comparing the results at Cleveland and Elizabethtown, we see that when the temperature was reduced by half, the ozone dose (mg O_3/mg C) has to be increased by 30% and Ct by 50% to achieve a similar level of inactivation.

Table 3.15
Ozone decay kinetic constants

Site	k_d (min-1)	Type of water	pH	Alkalinity (mg/L)	TOC (mg/L)	UV254 (cm-1)	Turbidity (NTU)	[O3]in* (mg/L)
Gwinnett	0.026	Filtered	6.95	17	3	0.0105	0.3025	0.49
	0.03	Filtered	7.12	17	2.6	0.01	0.605	1.03
	0.027	Filtered	7.09	17	2.4	0.01	0.376	1.56
	0.082	Raw	6.99	18	3	0.027	2.1	1.34
Cleveland	0.112	Raw	8.1	90	2.6	0.047	9.6	0.53
	0.078	Raw	8.1	90	2.5	0.028	12.9	1.06
	0.049	Raw	8.1	90	2.1	0.053	9.5	1.54
Elizabethtown	0.087	Raw	7.2	42	3.3	0.092	13	1.52
	0.110	Raw	7.2	39	4.9	0.08	5.90	1.34
	0.111	Raw	7.1	42	3.1	0.093	14	1.02
	0.075	Raw	6.5	42	3.1	0.093	14	0.97
	0.042	Raw	6.5	42	3.1	0.093	14	1.72
Los Angeles	0.07	Raw	8.13	109	1.94	0.053	1.6	1.09
	0.18	Raw	8.07	106	1.97	0.06	1.6	0.52
	0.053	Raw	6.5	106	1.97	0.06	1.6	0.66
	0.115	Raw	8.1	109	1.96	0.052	1.3	0.76
	0.033	Raw	6.5	109	1.96	0.052	1.3	1.6
UK – Chertsey	0.036	Filtered	7.3	230	2.6	0.14	0.12	1
	0.036	Filtered	7.3	236	2.9	0.135	0.1	0.6
	0.026	Filtered	7.3	234	2.6	0.133	0.13	1.4

* $[O_3]_{in}$: ozone concentration at the inlet of the contact columns

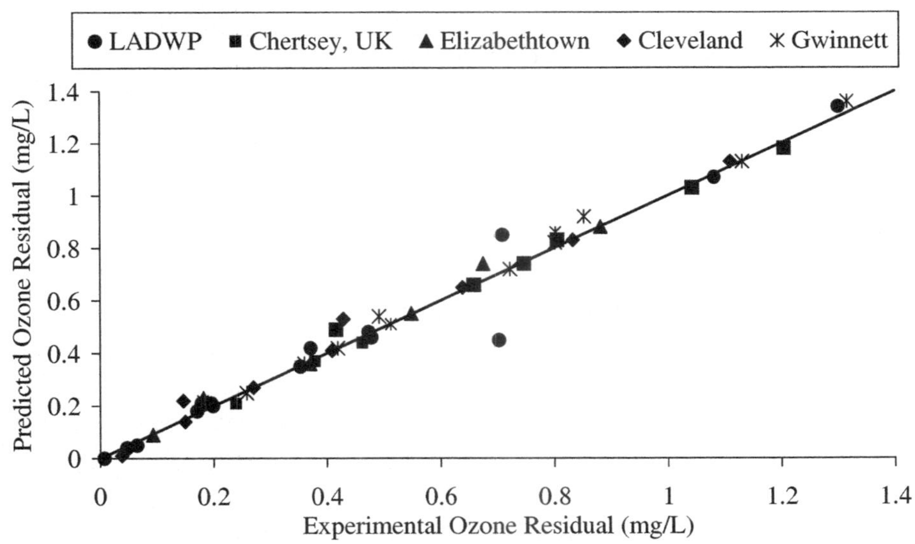

Figure 3.18 Ozone decay: predicted versus experimental

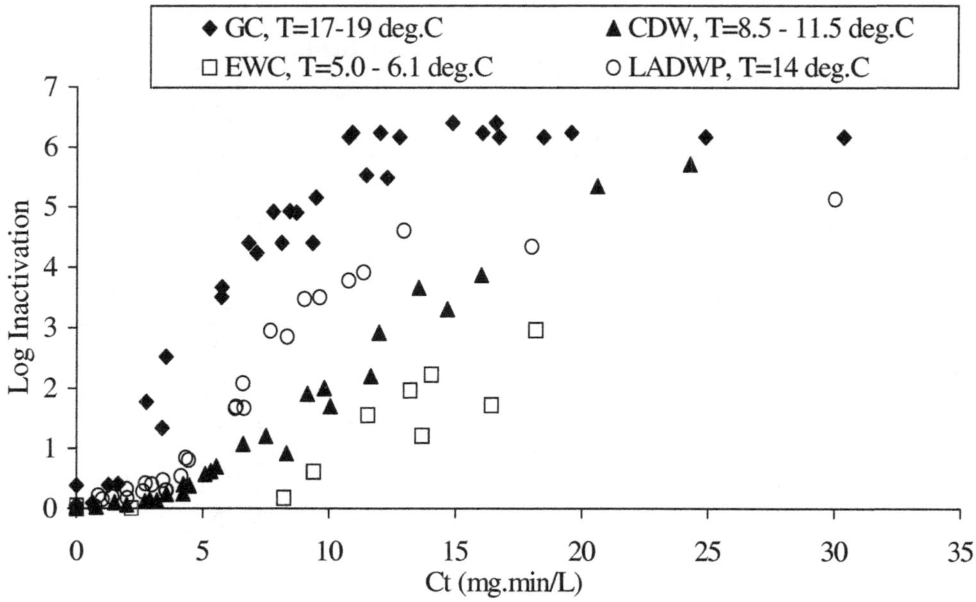

Figure 3.19 Inactivation of *Bacillus subtilis* as a function of Ct at different pilot-sites

Table 3.16
Cts required to achieve different levels of inactivation of *Bacillus subtilis*

Site	Water Temperature (°C)	Ct required for 2 Log inactivation	Ct required for 3 Log inactivation	Ct required for 4 Log inactivation	Ozone Dose* (mg O_3/mg C)
Chertsey	18	ED†	≈ 3.5	5	0.3
GC	17	3.0	5.0	6	0.3
LADWP	14	6.0	8.0	11.5	0.5
CWD	8.5-11.5	9.5	12.0	16.0	0.6
EWC	5-6.1	13.0	18.0	N.R.	0.8

N.R. = inactivation rate non-reachable after a contact time of 26 minutes

* Ozone dose required to achieve 3 Log inactivation at ambient pH after a contact time of 26 minutes.

† Erroneous data.

Bromate Formation

In waters where bromide was absent, potassium bromide was spiked to achieve the desired concentration. Bromate formation was studied as a function of various ozone residuals at the inlet of the contactor. During each trial, ozone residuals were measured in order to study the influence of the dissolved ozone profile on bromate formation. Pilot-scale testing at the MWTP was conducted using the SMU. Two strategies to control bromate formation were investigated over the different ozone conditions: 1) pH adjustment between 5.5 to 6.5, and 2) ammonia injection at two dosages (0.35 mg/L as N and 1.15 mg/L as N). During the period of testing, the bromide concentration in the outflow of the Actiflo process ranged between 110 and 120 µg/L.

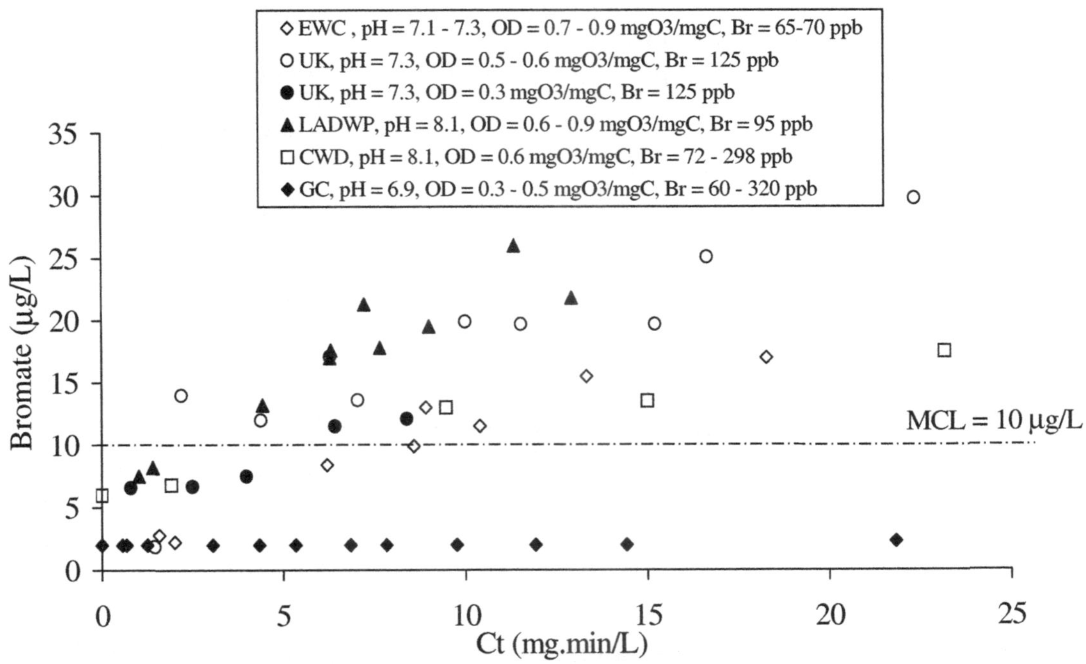

Figure 3.20 Bromate formation at ambient pH as a function of Ct, OD = ozone dose

At all the sites, for each trial, samples were taken along the contact columns for bromate analysis. Each sample was collected in duplicate and analyzed in triplicate.

Ambient pH

Figure 3.20 shows the bromate formed at different sites as a function of Ct at ambient pH. At LADWP, as the raw water contained high concentrations of bromide (90 to 100 μg/L), no spiking was necessary. At GC, for optimal ozone doses (0.3 to 0.5 mg O_3/mg C, Ct = 5 mg.min/L for 3 log inactivation of *Bacillus subtilis*) and varying spiked bromide concentrations (60 to 320 μg/L), the bromate formed was around 2 μg/L or less, even after 25.6 minutes of contact time. At higher ozone doses (0.6 mg O_3/mg C) more bromate was formed with low bromide concentration (60 μg/L) in the influent water underscoring the effect of ozone dose on bromate formation (data not shown). The results obtained at the CWTP, UK reinforces that observation as an ozone dose of 0.3 mg O_3/mg C lead to a bromate concentration less than the MCL after 26 minutes of contact time, whereas the bromate concentration exceeded 10 μg/L when applying higher ozone doses (>0.5 mg O_3/mgC).

At the other sites (CDW, EWC and LADWP), ozone doses higher than 0.5 mg O_3/mg C were applied in order to reach inactivation rates of *Bacillus subtilis* > 3 Log. Under such conditions the bromate concentration exceeded 10 μg/L. Higher Ct is due to the low temperature of these waters as compared to GC and CWTP water. It is also important to note that the pH of these waters were ≥ 7.1. Also, at CDW as the bromide concentration increased (Br^- = 72 to 298 μg/L, through spiking), the bromate formation also increased.

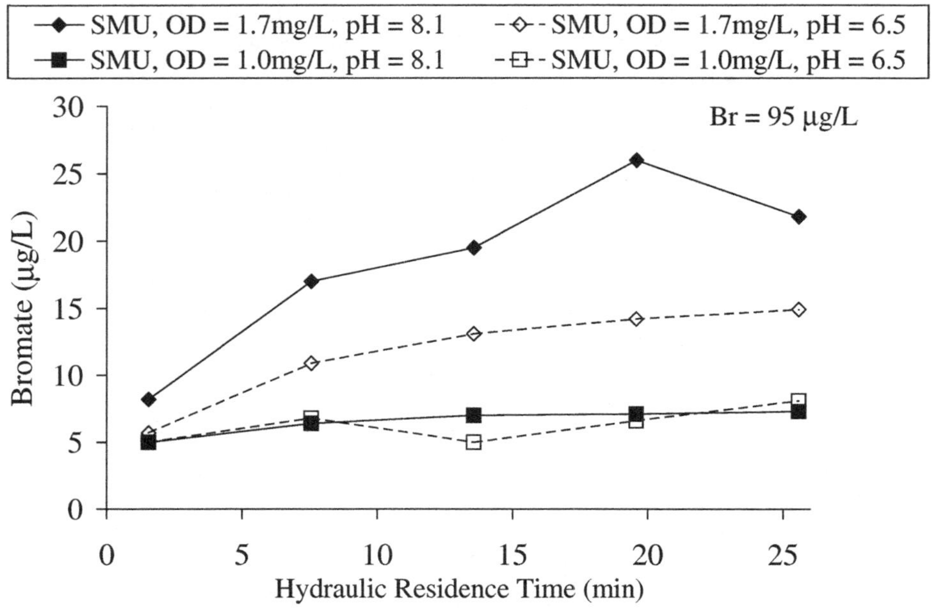

Figure 3.21 Bromate formation profiles along the SMU – Influence of pH – LADWP

Ambient pH and pH 6.5

Figure 3.21 depicts the bromate formed along the SMU for each trial in order to quantify the influence of pH on bromate formation as a function of hydraulic residence time at the LADWP. The highest ozone dose (1.7 mg/L, 1 mg O_3/ mg C) without pH adjustment lead to bromate exceeding the current MCL of 10 ppb. An ozone dose of 1.0 mg/L (0.5 mg O_3/ mg C) led to low bromate formation at a pH of 8.1. In fact, the bromate formation was lower at low ozone dose at ambient pH ($\approx$ 5 ppb) as compared to bromate formation ($\geq$ 10 ppb) at high ozone dose and lowered pH (6.5). An ozone dose of 0.5 mg O_3/ mg C and pH 6.5 led to a bromate formation of $\approx$ 5 ppb. The impact of pH seems to be not as significant as lowering the ozone dose (0.5 mg O_3/ mg C).

Lowering ozone dose to minimize bromate formation is a good strategy. However, one has to make sure that such an ozone dose will achieve adequate microbial inactivation. Figure 3.22 depicts the bromate concentration formed along the SMU for each run aimed at quantifying the influence of pH on bromate formation as a function of Ct formation (only LADWP results shown). Inactivation curves obtained through the SMU are also plotted in the same figure. An ozone dose of 1.0 mg/L (0.5 mg O_3/mg C) led to low bromate formation at a pH of 8.1 but after a contact time of 25.6 minutes the inactivation rate was lower than 0.5 Log of *Bacillus subtilis* spores. However, at the same applied ozone dose of 1.0 mg/L (0.5 mg O_3/mg C, Ct=8 mg.min/L) and a pH lowered to 6.5 allowed to obtain an 3.0 Log inactivation of *Bacillus subtilis* while being able to keep the bromate concentration well below MCL after a contact time of 25.6 minutes.

For the LADWP situation, applying an ozone dose of about 0.5 mg O_3/mg C through a static mixer dissolution system and lowering pH to 6.5 before ozonation seems to be the best strategy to meet the disinfection objectives and efficiently limit bromate formation. Moreover, lowering pH leads to a significant decrease in ozone decay along the contactor, therefore, for an

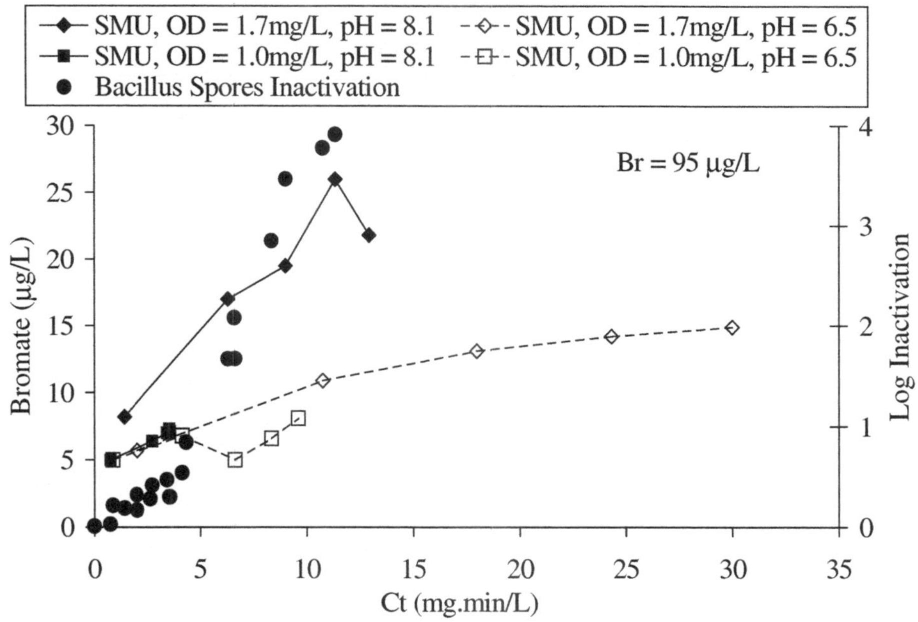

Figure 3.22 Bromate formation and log inactivation as a function of Ct – LADWP

identical applied ozone dose, Cts reached at the outlet of the contact columns with a pH adjusted to 6.5 are increased by twofold compared to those observed with ambient pH.

Additional tests were conducted in Tampa, Florida and the results on bromate formation showed that for all the samples analyzed, the bromate concentration was found under the detection limit (5 μg/L), whatever the operating conditions: i) ozone dose from 1.5 mg/L up to 3.5 mg/L (0.4 to 0.85 mg O_3/mg C), ii) pH between 5.5 and 6.5, iii) with or without ammonia addition. These results confirm the observations noted in LADWP and GC. In fact, even with high ozone doses (0.85 mg O_3/mg C), the low pH (6.5 maximum) process prevented excess bromate formation.

Impact of Alkalinity on Bromate Formation

Song et al. (1997) developed an overall empirical model on bromate formation. In this model, an increase in alkalinity level was found to result in increased amounts of BrO_3^-. In order to verify this impact of alkalinity on bromate formation, four raw drinking waters sources were compared. Each type of water was selected in order to represent an element of the first row of the 3*3 matrix for enhanced coagulation as prescribed in the U.S. Environmental Protection Agency's Disinfectant/Disinfection By-Products Rule. This matrix is presented in Table 3.17. The results obtained are presented in Figure 3.23.

As demonstrated previously, pH and O_3/DOC ratio have a significant impact on bromate formation. Trials were selected from the pilot study database in order to compare different alkalinity levels with approximately the same pH (6.5 to 7.3) and O_3/DOC ratio (0.43 to 0.58 mg O_3/mgC). Results clearly confirm the model of Song et al. Higher alkalinity levels lead to increased bromate formation as a function of Ct. Low pH and low alkalinity waters (GC, MWTP) led to low bromate formation.

Table 3.17
Enhanced coagulation matrix

TOC (mg/L)	Alkalinity, mg/L as $CaCO_3$		
	0 to 60	60 to 120	> 120
2-4	Gwinnett County, GA MWTP, FL EWC, NJ	Los Angeles, CA Cleveland, OH	Chertsey, UK

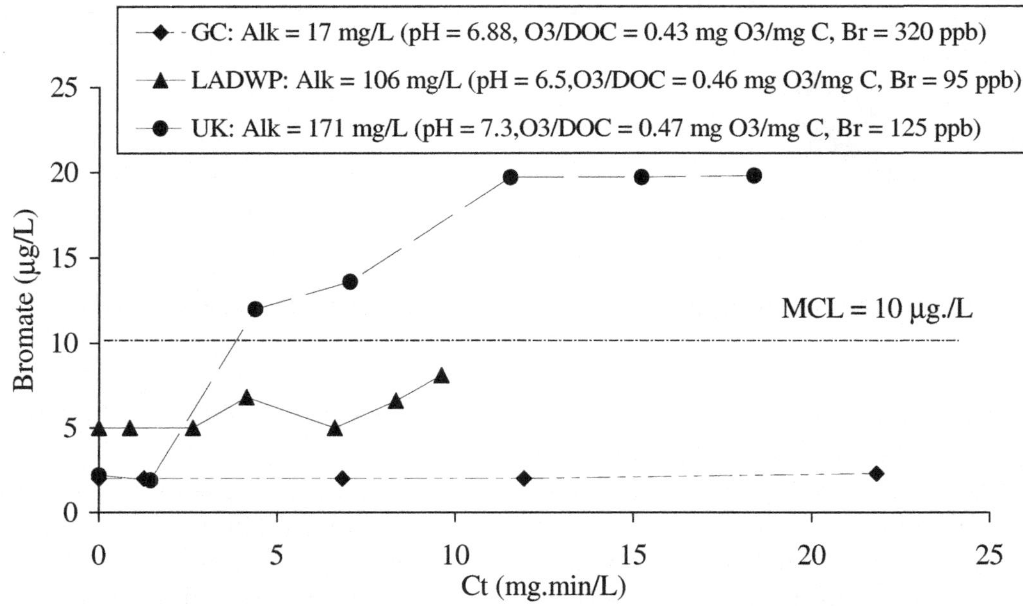

Figure 3.23 Bromate formation versus Ct for water with different alkalinities

Side-by-Side Studies With SMU and FBDU

Pilot studies were conducted side-by-side with the static mixer pilot unit (SMU) and the conventional fine bubble diffusion pilot unit (FBDU) to evaluate ozone dissolution performance, microbial inactivation, and bromate formation at LADWP and EWC. Side-by-side comparison tests were conducted by applying similar ozone doses and feeding the same raw water to both units at LADWP and EWC.

Ozone Profiling

Table 3.18 shows the operating conditions and the Ct achieved during these runs at LADWP. CTs during the same runs are shown in Figure 3.24. During each run, transfer efficiencies were measured. Under the tested operating conditions, the FBDU provided a transfer efficiency of 97% while a transfer efficiency of 94% to 97% was obtained with the SMU (G/L = 1.5-2.7%). Under the same applied ozone dose, higher ozone residuals were obtained in a short interval of time at the outlet of the dissolution system for SMU when compared to the FBDU.

Table 3.18
Operating conditions and Ct obtained for each test – LADWP

Operating Conditions	Static Mixer Unit (SMU)	Fine Bubble Diffuser Unit (FBDU)
Low Ozone Dose OD = 1.0 mg/L	Ozone Residual = 0.5 mg/L* $CT_{19\ min}$ = 3.5 mg.min/L† Overall Ct = 3.6 mg.min/L‡	Ozone Residual = 0.3 mg/L* Overall Ct = 1.5 mg.min/L‡
Medium Ozone Dose OD = 1.2 mg/L	Ozone Residual = 0.8 mg/L* $CT_{19\ min}$ = 6.3 mg.min/L† Overall Ct = 7.7 mg.min/L‡	Ozone Residual = 0.3 mg/L* Overall Ct = 1.3 mg.min/L‡
High Ozone Dose OD = 1.7 mg/L	Ozone Residual = 1.1 mg/L* $CT_{19\ min}$ = 11.3 mg.min/L† Overall Ct = 12.9 mg.min/L‡	Ozone Residual = 1.3 mg/L* Overall Ct = 11.2 mg.min/L‡
High Ozone Dose OD = 1.7 mg/L With pH lowered to 6.5	Ozone Residual = 1.6 mg/L* $CT_{19\ min}$ = 18.0 mg.min/L† Overall Ct = 30.0 mg.min/L‡	Ozone Residual = 1.2 mg/L* Overall Ct = 10.6 mg.min/L‡
Low Ozone Dose OD = 1.0 mg/L With pH lowered to 6.5	Ozone Residual = 0.7 mg/L* $CT_{19\ min}$ = 8.3 mg.min/L† Overall Ct = 9.6 mg.min/L‡	Ozone Residual = 0.5 mg/L* Overall Ct = 3.6 mg.min/L‡

* Ozone residual at the outlet of the dissolution system (outlet of the gas-liquid separator for the SMU, and outlet of the first column for the FBDU).

† Ct reached at the outlet of the third contact column of the SMU. At that point, the contact time is the same as the overall contact time provided by the FBDU.

‡ Overall Ct calculated at the outlet of the contactors.

The main difference is that the SMU dissolution system only provides a 1.5 minutes contact time whereas the one for the FBDU requires a 6.2 minutes contact time. Intense mixing and ozone transfer through the static mixer elements leads to a much faster ozone dissolution leading to higher ozone residuals than the FBDU. This faster ozone dissolution via the SMU has, therefore, a large influence on the ozone profile and the Cts obtained along the contactors located downstream of the dissolution system.

On operating both units with the same applied ozone dose and owing to its faster ozone dissolution performances, Cts along the SMU were systematically higher (1.7 to 2.3 times) than those provided through the FBDU of equivalent contact time. For example, if the targeted overall Ct of the ozonation treatment were 5 mg.min/L, then, depending on the ozone dissolution method used, the contact times to be provided would be as shown in Table 3.19.

As the contact time required to achieve comparable Cts is lowered with a static mixer, the overall footprint of the ozone process is significantly reduced. Lowering pH leads to a significant decrease in ozone decay along the contactor. Hence, for an identical applied ozone dose, Cts reached at the outlet of the contact columns with an adjusted pH of 6.5 are doubled compared to those observed with natural pH (8.1).

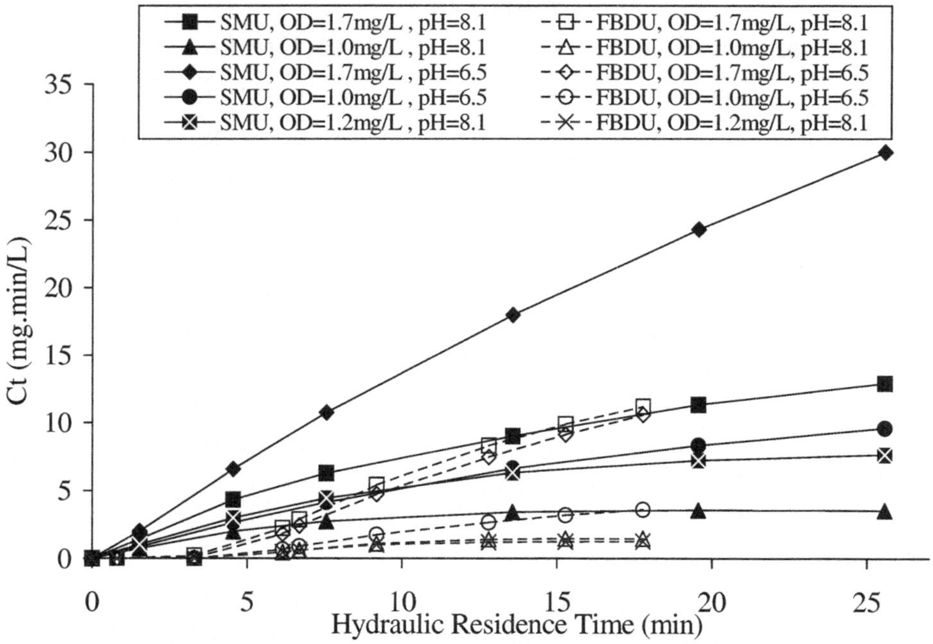

Figure 3.24 Cts profiles along SMU and FBDU obtained with similar applied ozone doses – LADWP

Table 3.19
Contact times to reach a Ct of 5 mg.min/L

	Static Mixer Unit	Fine Bubble Diffuser Unit
Ozone Dose = 1.7 mg/L ; pH = 8.1	5 min.	8 min.
Ozone Dose = 1.7 mg/L ; pH = 6.5	3 min.	10 min.
Ozone Dose = 1.0 mg/L ; pH = 8.1	N.R.*	N.R.*
Ozone Dose = 1.0 mg/L ; pH = 6.5	10 min.	> 18 min.

* Ct non reachable.

At the EWC, results similar to LADWP were obtained with respect to CTs obtained downstream of the contactors. Cts provided during each test are depicted in Figure 3.25. On operating both units with the same applied ozone dose and owing to its faster ozone dissolution performances, Cts along the SMU were systematically higher (1.2-1.5 times) than those provided through the FBDU of equivalent contact time. For example, if the targeted overall Ct of the ozone treatment is 10 mg.min/L, then, depending on the ozone dissolution technology used, the contact time to be provided would be as shown in Table 3.20. For both units, the dissolution system was taken into account within the calculation of the Ct values.

Disinfection Performance

Figure 3.26 depicts the inactivation curves obtained with the SMU and FBDU at similar applied ozone doses, and at two different pHs. Due to the higher Cts reached when operating with

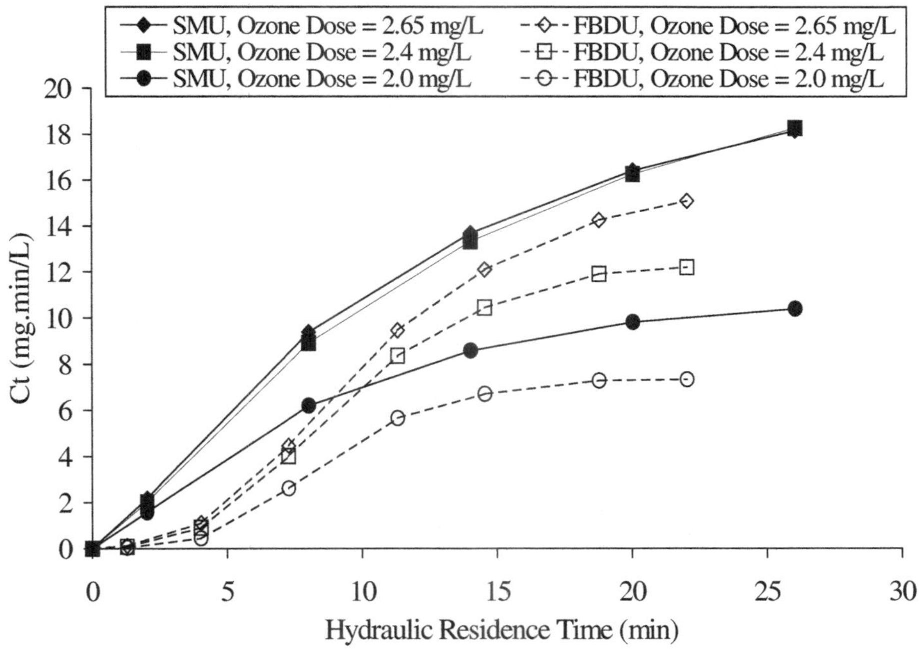

Figure 3.25 Cts profiles along both units obtained with similar applied ozone doses – EWC

Table 3.20
Contact times to reach a Ct of 10 mg.min/L-EWC

	Static Mixer Unit	Fine Bubble Diffuser Unit
Ozone Dose = 2.65 mg/L	9 min	12 min
Ozone Dose = 2.4 mg/L	10 min	14 min
Ozone Dose = 2.0 mg/L	20 min	> 25 min

a lowered pH (6.5); the inactivation rates at the outlet of the contactor are increased for both the units. The inactivation results show that the SMU provides enhanced inactivation of 0.4 to 2.9 Log compared to the FBDU under similar applied ozone doses, water quality conditions, and equivalent contact times. This improvement is due to a consequence of the higher Cts reached along the SMU. As an example, a maximum Ct of 8 mg.min/L achieved in an SMU resulted in a 3 Log inactivation of *Bacillus subtilis* utilizing an ozone dose of 1.2 mg/L. Under similar conditions, a maximum Ct of 1.3 mg.min/L was achieved in the FBDU, which resulted in a 0.2 Log inactivation of *Bacillus subtilis*.

At the EWC, due to the cold water temperature, a high and variable turbidity of the water to be treated, ozone doses (greater than 2.0 mg/L) were applied in order to reach a significant level of inactivation. Results similar to LADWP were obtained (Figure 3.27). An ozone dose of 0.8 mg O_3/mg C (Ct=18 mg.min/L) was necessary to achieve a 3 Log inactivation due to the low temperature of the Raritan Millstone river.

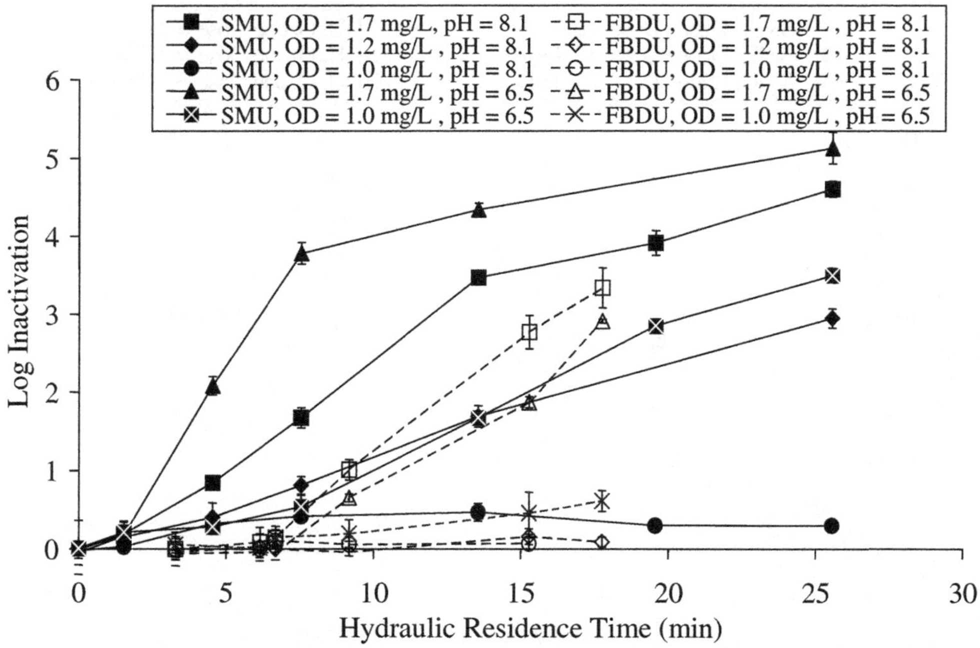

Figure 3.26 Inactivation of *Bacillus subtilis* as a function of HRT for SMU and FBDU – LADWP

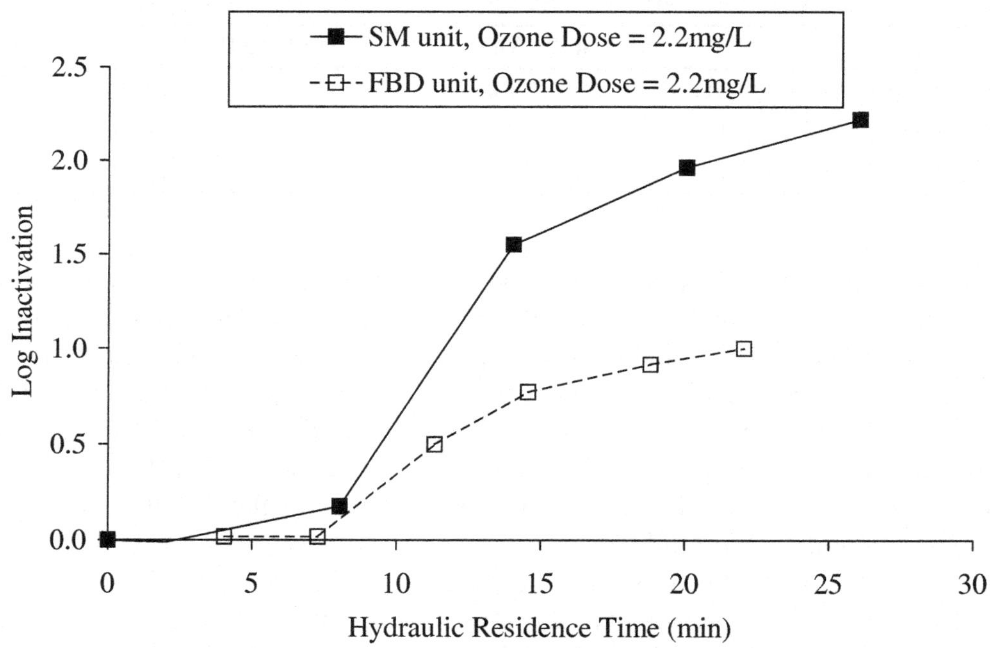

Figure 3.27 Inactivation of *Bacillus subtilis* as a function of HRT for SMU and FBDU – EWC

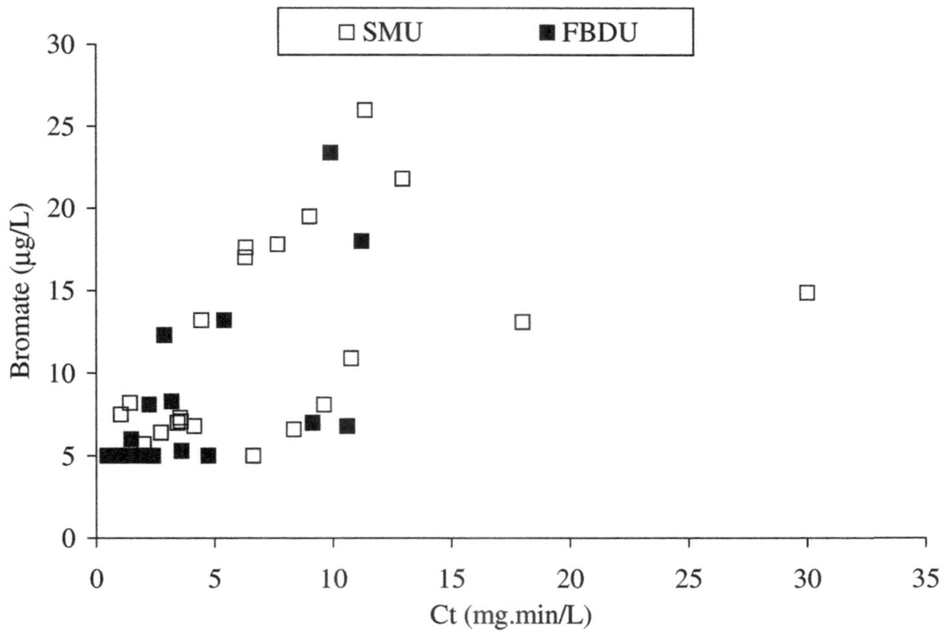

Figure 3.28 Comparison of bromate formation versus Ct between SMU and FBDU in LADWP

Bromate Formation

No significant difference in bromate formation was observed between SMU and FBDU at the LADWP (Figure 3.28). At the EWC, bromide was spiked at a concentration of 65 µg/L, and pH of the raw water during these test runs was between 7.1 and 7.3. Bromate formed as a function of Ct is shown in Figure 3.29. Bromate formed was higher than the MCL for Cts greater than 10. Under an ozone dose of 2.4 mg/L (1.0 mg O_3/mg C), bromate concentration exceeds 10 µg/L for both units. For the lowest ozone dose (2.0 mg/L, 0.7 mg O_3/mg C), bromate formation is significantly minimized along the SMU and the FBDU. At ambient pH, the FBDU formed slightly more bromate than the SMU for similar Cts.

Two sets of tests were carried out on the SMU to quantify the influence of pH (ambient and 6.5) on bromate formation through ozone treatment. The results showed that bromate formation was reduced to 5 ppb or less as a consequence of lowering pH to 6.5 (data not shown). Also, the ozone decomposition along the ozone contactors was lowered as well, which results in higher Cts.

Pesticide Oxidation Performance

The tested pesticide was metolachlor, which is a commonly-used herbicide in the United States. During two runs, metolachlor was injected to reach an initial concentration of about 5 ppb. Oxidation results show that the static mixer unit provides total pesticide destruction after a contact time of 26 minutes (Figure 3.30). Intense mixing through the static mixer provides for good oxidation (>40%) after a contact time of less than a second.

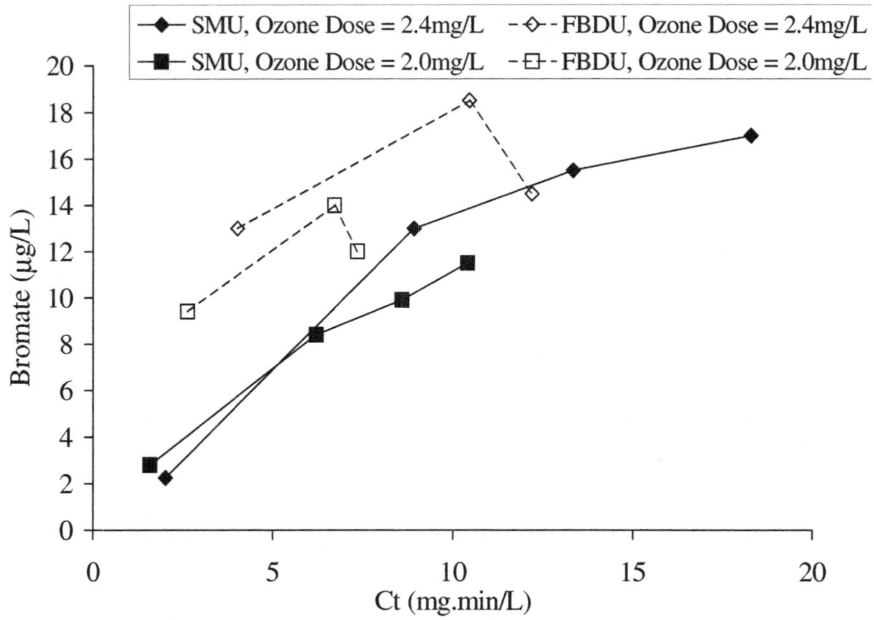

Figure 3.29 Bromate formation through both units as a function of Ct – EWC

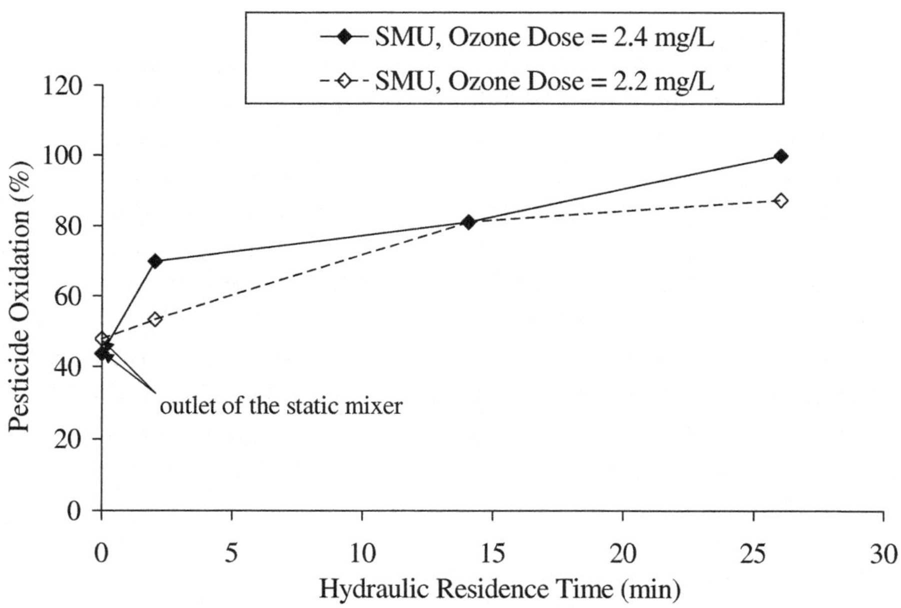

Figure 3.30 Metolachlor oxidation results for the static mixer unit

Table 3.21
Water quality parameters – Inter-ozonation inflow

Parameter	Value
Water Temperature (°C)	18
pH	7.3
Alkalinity (mg $CaCO_3$/L)	230 to 236
UV_{254} (cm^{-1})	0.133 to 0.140
DOC (mg C/L)	2.5 to 2.7
TOC (mg C/L)	2.6 to 2.9
Bromide (μg/L)	125
Turbidity (NTU)	0.10 to 0.13

Table 3.22
Static mixer operating parameters at full-scale at the inter-ozonation of CWTP

Parameter	Value
Configuration	Vertical Down
Type of Ozone Application	Direct Injection
Liquid Flowrate (m^3/h)	630 to 1355
Contact Time (min)	3.4 to 8
Applied Ozone Dose (mg/L)	0.8 to 1.0
Applied Ozone Dose (mg O_3/mg C)	0.3 to 0.4

Side-by-Side Studies With SMU and a Full-Scale Static Mixer at CWTP, UK

The final phase of this task, involved comparing static mixer performances at pilot- and full-scale in the UK. For full-scale demonstrations, Vivendi Water Partnership (VWP) in the United Kingdom were utility contributors and provided full-scale operation data on different ozone-static mixer installations. More details are provided in the next Chapter. The static mixer pilot unit was operated side by side to the full-scale plant i.e. Chertsey Water Treatment Plant, (CWTP) UK, which also employs static mixers for ozonation. The objectives were to compare ozone dissolution performances, microbial inactivation, and bromate formation.

At the Chertsey WTP, testing was carried out at the intermediate ozonation stage. The inflow at that location of the process train is a blend of groundwater (Abbeymeads), pre-ozonated and passed through rapid gravity filters, and surface water (River Thames) pre-ozonated, treated through rapid gravity filters and slow sand filters. Water quality parameters of the incoming water to the ozone process employing static mixers at the Chertsey WTP during the period of testing are shown in Table 3.21.

Table 3.22 summarizes the main operating parameters of the static mixer used at the intermediate ozonation stage of the CWTP.

The performance evaluation was based on the following criteria: ozone dissolution performance, disinfection effectiveness and bromate formation.

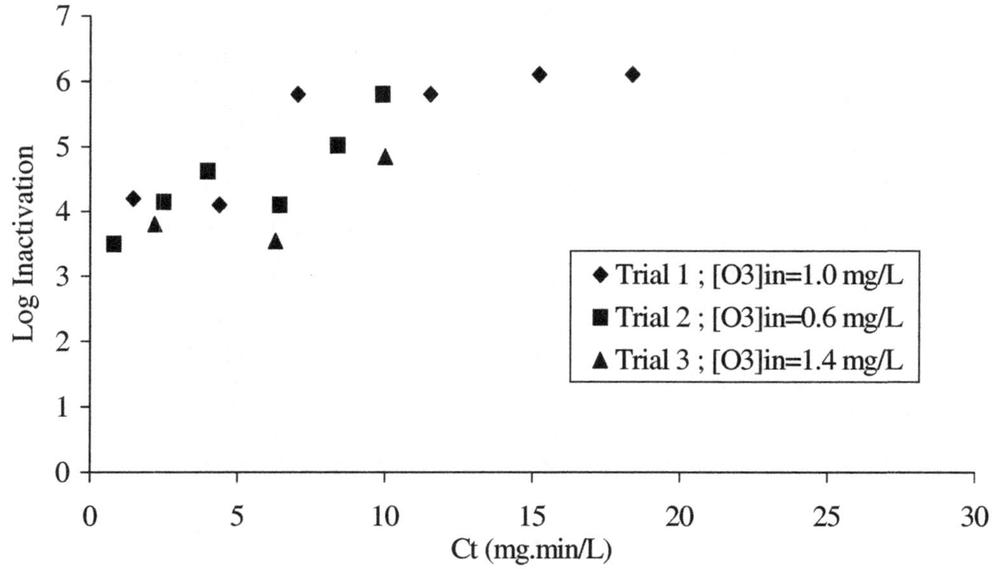

Figure 3.31 ***Bacillus subtilis*** **spores inactivation along the SMU at pilot-scale at CWTP, UK**

Transfer Efficiency

Dissolved ozone concentrations were measured along the contactor for the three microorganism reduction assays. Ozone transfer efficiency through the dissolution system was determined and was found to be >95% for all the trials. Operating conditions have little influence on ozone transfer performance over this range of ozone dose (0.8 mg/L to 1.7 mg/L). In the full-scale plant transfer efficiency superior to 95% was achieved at all times.

Disinfection Performance

Bacillus subtilis spores were spiked at the inflow of the SMU to get an initial *Bacillus subtilis* concentration of 10^4/mL. Samples were taken along the SMU, and *Bacillus subtilis* were enumerated to determine the inactivation rates for the various trials. Figure 3.31 depicts the inactivation performance as a function of Ct for each test. Due to the warm temperature (18°C), *Bacillus subtilis* inactivation was enhanced. For a Ct of > 5 mg.min/L, 4 Log inactivation was observed. This compares favorably with the results obtained at Gwinnett County (temperature = 17°C, CT required for 4 log inactivation = 6 mg.min/L, Table 3.16). At low Cts, < 3 mg.min/L, 4 log inactivation was observed which was contrary to all previous results obtained. In comparison, at a similar Ct, inactivation was only about 1.5 log-units at Gwinnet County. Obviously the results from Chertsey at lower Cts were in error and not used (Table 3.16). For the full-scale ozone system, lower inactivations of *Bacillus subtilis* (0.2 to 0.7 Log) were observed. This can be explained by the fact that the concentration of *Bacillus subtilis* in the filtered water entering the intermediate-ozonation step was relatively low (20 to 330/100ml).

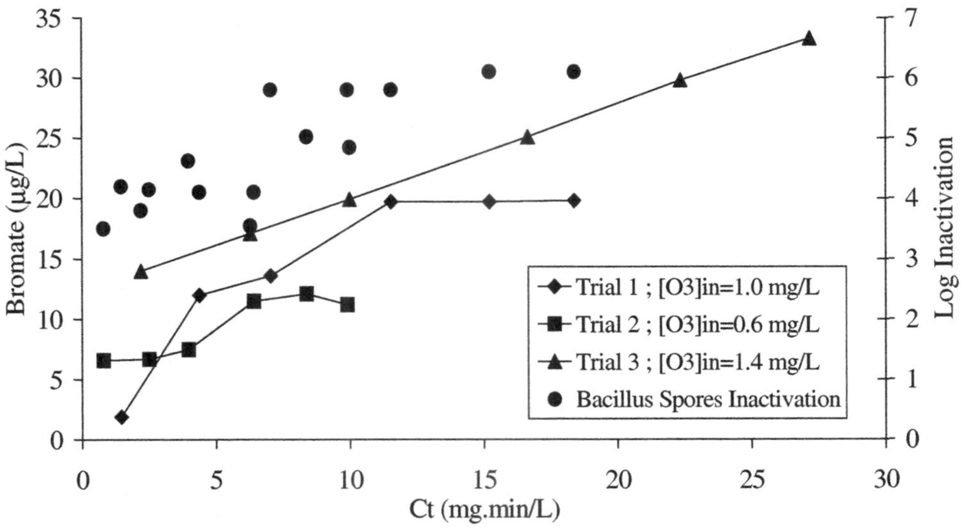

Figure 3.32 Bromate formation and log inactivation as a function of Ct at pilot-scale at CWTP, UK

Bromate Formation

Water samples during each test were collected along the pilot unit for bromide and bromate analysis. Due to the high naturally-occurring bromide concentration (123 µg/L to 126 µg/L) in the filtered water entering the pilot unit, no spiking was required.

Figure 3.32 depicts the bromate formation as a function of Ct. The log inactivation of *Bacillus subtilis* is also shown in the same graph. Bromate formation along the unit was found to be largely dependent on the applied ozone dose. For the lowest applied ozone dose (0.8 mg/L, 0.3 mg O_3/mg C), the bromate concentration was ≤ 10 µg/L, whereas for the two highest applied ozone doses (1.2 mg/L or 0.5 mg O_3/mg C and 1.7 mg/L or 0.7 mg O_3/mg C) the bromate concentration exceeds 10 µg/L after a contact time of 4 minutes. The pH of the water tested was 7.3. Time did not permit us to run any tests with lowered pH.

For a water temperature of 18°C, a Ct of 5 mg.min/L is sufficient to reach more than 4 Log inactivation of *Bacillus subtilis* spores and to limit the bromate concentration under 10 µg/L. The operating conditions adjusted for Trial 2 (applied ozone dose = 0.8 mg/L, 0.3 mg O_3/mg C) enabled to attain both objectives as long as the overall contact time was lower than 15 minutes. In the full-scale plant, the ozone dose varied between 0.3 and 0.4 mg O_3/mg C and, since the contact time varies between 3.4 to 8 minutes for intermediate ozonation (see Table 3.22), bromate formation was always lower than 10 µg/L (Figure 3.33). The contact time in this plant varies between 1.1 to 3.2 minutes for pre-ozonation.

Comparison of Ct Values

The Ct values for *Bacillus subtilis* and *Cryptosporidium parvum* obtained during laboratory-scale, pilot-scale, modeling and from literature are shown in Table 3.23 for high (22°C), moderate (10 to 18°C) and low temperature (< 5°C). Comparing Ct values of *Bacillus*

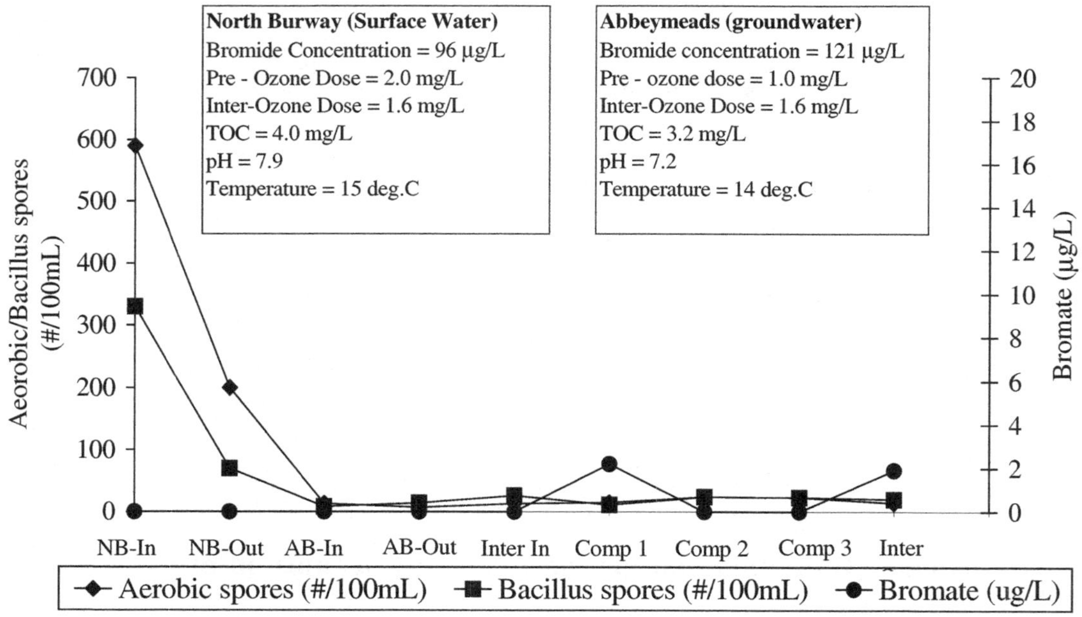

Figure 3.33 Effect of ozone on aerobic spores, *Bacillus subtilis*, and bromate formation along CWTP, UK (May 2000)

subtilis (batch) and *Cryptosporidium parvum* (modeling) at 12°C/3°C and Ct of *Bacillus subtilis* (pilot data, CWD) *and Cryptosporidium parvum* (Oppenheimer et al. 1999) at 10°C we see that, *Bacillus subtilis* is inactivated more readily than *Cryptosporidium parvum*. Hence, at these temperatures (12°C to 3°C), 2-log inactivation of *Cryptosporidium parvum* can be achieved at 2 to 2.3 times the equivalent *Bacillus subtilis* Ct value.

Comparing Ct values of *Cryptosporidium* (Finch) at high and low temperatures with the Cts obtained from this study either thru lab work or modeling we find that Ct required for 2 log inactivation is reduced by 42-52%. Similar trends can be seen while comparing Ct values derived in natural waters (Oppenheimer et al. 1999) and this work. This could have serious implications on the development of Ct tables for inactivation of *Cryptosporidium parvum*.

Disinfection studies can be complex to design and construct, especially when pathogenic protozoa (i.e. *Cryptosporidium* and *Giardia*) are included. In such studies and other performance-based testing, where quantifying the inactivation of target microorganisms (e.g. *Cryptosporidium*) are required, *Bacillus subtilis* can serve as a useful surrogate. *Bacillus subtilis* spores are also useful for evaluation of the hydrodynamic efficiency of alternative ozone contactor designs, as they were found to be relatively resistant to ozone over a large temperature and pH range.

SUMMARY

The French pilot studies showed that while using a static mixer configuration of 4 mixing elements, the optimal transfer efficiency is reached with a low G/L ratio. An increase in liquid velocity is also favorable to optimize this parameter. Nevertheless, a plateau at 0.8 m/s is noticed with a G/L ratio less than 3%. Beyond this velocity, there is no improvement of transfer efficiency. This velocity corresponds to a pressure drop of 10 KPa (1.45 psi). The ozone dose and the ozone residual at the outlet of the static mixer have little influence on the transfer efficiency

Table 3.23
Comparison of Ct values from this work and literature

Microorganism	Source	Temperature(°C)	Ct for 2 log inactivation
	High Temperature: 22°C		
Cryptosporidium	Finch, 1999	22	4.4
Cryptosporidium	Oppenheimer et al., 1999	22	2.9
Cryptosporidium	This study, RFS lab data	22	2.1
Bacillus subtilis	This study, batch lab data	22	2.5
	Moderate Temperature : 10 to 18°C		
Cryptosporidium	Finch, 1999	13	11
Cryptosporidium	Oppenheimer et al., 1999	10	18
Cryptosporidium	This work, modeling	12	11
Cryptosporidium	This work, modeling	16	7
Bacillus subtilis	This work, batch lab data	16	7
Bacillus subtilis	This work, batch lab data	12	5
Bacillus subtilis	This work, pilot data from GC	18	3
Bacillus subtilis	This work, pilot data from LADWP	14	6
Bacillus subtilis	This work, pilot data from CWD	10	9.5
	Low Temperature : <5°C		
Cryptosporidium	Finch, 1999	1	40
Cryptosporidium	Oppenheimer et al., 1999	3	52
Cryptosporidium	This work, modeling	3	23
Bacillus subtilis	This work, batch lab data	3	10
Bacillus subtilis	This work, pilot data from EWC	5	13

compared with the first two variables. With this configuration, the transfer efficiencies that can be achieved are greater than 90% with a G/L ratio close to 1.5% and a liquid velocity greater than 0.75 m/s.

The residence time distribution of the pilot–unit used for US and UK studies was characteristic of a plug flow with dispersion type hydraulic regime with no short-circuiting. This hydraulic characteristic was useful to achieve high levels of microbial inactivation in disinfection experiments. For the waters tested, a G/L ratio of 0.6 to 1.5 % resulted in the highest transfer efficiency (96% to 99%).

Pilot studies were conducted at the LADWP and EWC where the SMU was compared side-by-side with a fine bubble diffusion pilot unit FBDU. Under the same applied ozone dose, higher ozone residuals were obtained in a short interval of time at the outlet of the dissolution system for SMU when compared to the FBDU. The main difference is that the SMU dissolution system only provides a 1.5 minutes contact time whereas the one for the FBDU requires a 6.2 minutes contact time. Intense mixing and ozone transfer through the static mixer elements leads to a much faster ozone dissolution leading to higher ozone residuals than the FBDU. This faster ozone dissolution via the SMU has, therefore, a large influence on the ozone profile and the Cts obtained along the contactors located downstream of the dissolution system. On operating both units with the same applied ozone dose and owing to its faster ozone dissolution performances, Ct values along the SMU were systematically higher (1.2-2.3 times) than those provided through the FBDU of equivalent contact time. Inactivation data obtained with *Bacillus subtilis* at LADWP and EWC showed that the SMU provided enhanced inactivation (0.4 to 2.9 Log more than FBDU) under similar applied ozone doses, water quality conditions, and equivalent contact times. This improvement is due to the higher Cts reached along the SMU. At the LADWP (water temperature=14°C), operating the SMU with an applied ozone of 1.0 mg/L (0.5 mg O_3/mg C, CT=9.6) and a pH lowered to 6.5 allowed to obtain an 3.5 Log inactivation of *Bacillus subtilis* while being able to keep the bromate concentration less than the MCL after a 25.6 minutes contact time. The observed inactivation was higher at an adjusted pH of 6.5 than the natural water pH due to a higher Ct obtained at a lower pH. Also, bromate formation was minimized at a lower pH (6.5). At EWC, the effect of water temperature (5°C) was more pronounced and an applied ozone dose of 2.7 mg/L (0.8 mg O_3/mg C, Ct=18) was necessary to achieve a 3 Log inactivation of *Bacillus subtilis* at a pH of 7.1.

Ozone dissolution performances, microbial inactivation, and bromate formation at pilot- and full-scale were compared side-by-side at the Chertsey WTP in the UK. It was found that the transfer efficiencies of both the pilot and full-scale systems were comparable. Inactivation data obtained with the pilot unit showed that > 4 Log inactivation of *Bacillus subtilis* was achieved for a Ct of 5 mg.min/L (at a low ozone dose of 0.3 mg O_3/mg C). In the full-scale plant lower inactivations (0.2 to 0.7 Log inactivation of *Bacillus subtilis*) were observed due to low concentrations of the microorganisms in the water. It was found that bromate formation along the pilot unit was found to be largely dependent on the applied ozone dose. For the lowest applied ozone dose (0.8 mg/L), the bromate concentration was ≤ 10 μg/L, whereas for the two applied ozone doses (1.2 mg/L and 1.7 mg/L) the bromate concentration exceeded 10 μg/L. For the full-scale intermediate-ozonation system, the ozone dose varied between 0.3 to 0.4 mg O_3/mg C and bromate formation was always lower than 10 μg/L, in agreement with pilot-scale results.

A comparison of the results from pilot-scale studies conducted at the various sites in terms of ozone dissolution performances, microbial inactivation and bromate formation showed that higher ozone residuals were obtained in a short interval of time at the outlet of the static mixer due to intense mixing and ozone transfer through the static mixer elements. Enhanced microbial inactivation was observed due to a consequence of the higher Cts reached along the static mixer unit. The results also confirmed that water quality parameters (temperature and pH among other factors) influence disinfection efficiency and bromate formation. The effect of temperature on inactivation of *Bacillus subtilis* with the pilot-scale unit confirmed the results obtained with the laboratory-scale unit and the well-stirred batch reactor. To achieve comparable inactivation, higher Cts are required at cold temperatures. The results underscored the importance of ozone dose and pH on bromate formation. The results also showed that bromate formation increases rapidly as a function of Ct for high alkalinity waters. In general, applying low ozone doses (≤ 0.5 mg O_3/mg C) through a static mixer ozone dissolution system integrated with a well designed downstream contactor (plug flow) combined with a pH (≤ 7.0) can provide adequate microbial inactivation while keeping bromate formation below the current MCL (10 ppb). For some waters, it may be necessary to depress the pH to 6.5 to achieve similar results.

The Ct values for *Bacillus subtilis* and *Cryptosporidium parvum* obtained during laboratory-scale, pilot-scale studies, modeling and from literature were compared at different temperatures. The comparison indicated that at higher temperatures (≥ 16°C) the Ct requirements for similar levels of log inactivation of both microorganisms are comparable. At temperatures 12°C to 3°C, *Bacillus subtilis* is inactivated more readily than *Cryptosporidium parvum*. The comparison of Ct values at these temperatures showed that 2-log inactivation of *Cryptosporidium parvum* can be achieved at 2-2.3 times the equivalent *Bacillus subtilis* Ct value.

The SWTR Guidance Manual provides technical guidance for the design of surface water treatment facilities (1990). The guidance manual focuses on the inactivation of *Giardia* and viruses, and Ct values for ozone inactivation of *Giardia* are provided in Appendix O. No Ct values are specified for the inactivation of *Cryptosporidium*; however, a commonly used method for determining *Cryptosporidium* inactivation has been to apply a linear multiplier to the tabulated *Giardia* Ct values to determine the *Cryptosporidium* Ct. In the absence of guidelines, ozone system designers have typically used multipliers ranging from 5 to 20 times *Giardia* Cts to determine Cryptosporidium inactivation Cts. In comparison, the potential for *Bacillus subtilis* to serve as a surrogate for *Cryptosporidium parvum* appears to be much better.

Disinfection studies can be complex to design and construct, especially when pathogenic protozoa (i.e. *Cryptosporidium* and *Giardia*) are included. In such studies and other performance-based testing, where quantifying the inactivation of target microorganisms (e.g. *Cryptosporidium*) are required, *Bacillus subtilis* will serve as a very useful surrogate.

On comparing Ct values for 2 log inactivation of *Cryptosporidium* (Finch, unpublished) at high and low temperatures with the Cts obtained from this study either thru lab work or modeling it was found that Ct required was reduced by 42-52%. Similar trends were seen while comparing Ct values derived in natural waters (Oppenheimer, 1999) and this work. This could have serious implications on the development of Ct tables for inactivation of *Cryptosporidium parvum*.

The overall results have important implications in the design of ozone contactors with static mixers for full-scale plants. The utilization of static mixers as an ozone dissolution technology can lead to reduction in the contactor size downstream of the ozone dissolution system as targeted inactivation efficiencies are reached more quickly with an optimized Ct as compared to conventional technologies. As the hydraulic residence time of the static mixer is very small (few seconds), the overall footprint of the ozonation process is also significantly reduced.

CHAPTER 4
FULL-SCALE STUDIES

INTRODUCTION

The data collected from this part of the project was used to support and confirm the results from the laboratory- and pilot-scale tests and to verify the hydraulic and kinetic models. The goals of the full-scale program was to study microbial inactivation, pesticide/micropollutant destruction and bromate formation under typical operating conditions at four water treatment plants (WTPs) fitted with static mixers for ozonation. Information was also collected, for comparative purposes, from a fifth WTP which utilizes turbine diffusers and porous plate diffusers for the pre- and inter-ozonation stages respectively. Data was used to verify hydrodynamic and mass transfer models. To satisfy the goals, the full-scale field testing was undertaken at the following five sites, all managed by Vivendi Water Partnership:

Chertsey WTP	Walton WTP	Iver WTP
Clay Lane WTP	Egham WTP	

The details of process trains at each of the above sites are included in Appendix B. The WTPs were selected to include a range of water quality types. Iver WTP, which does not use static mixers, was included to allow a comparison between static mixers and more conventional diffusers. The parameters analyzed were the same as those included in the laboratory and pilot-scale phase of the program to allow meaningful comparisons to be made between the different aspects of the study. The inactivation of *Bacillus subtilis* by ozone formed the basis for comparing the three scales of process (bench-, pilot- and full-scale) examined in this project.

MATERIALS AND METHODS

The materials and methods used for the measurements of ozone, bromide/bromate, and culturing/enumeration of *Bacillus subtilis* were the same as described in the previous Chapters.

Water Treatment Plants

Five sites were selected; three are surface water treatment plants withdrawing from the River Thames (Egham, Walton and Iver). One site uses a combination of surface water and ground water (North Burway for the surface water and Abbeymeads for the ground water at Chertsey) and one is a ground water site (Clay Lane). The plants were selected so as to give a great range of water quality types as possible.

Static Mixers

Four of the sites (Egham, Walton, Chertsey and Clay Lane) use static mixers for their ozonation process whilst one (Iver) uses turbines and conventional diffusers, enabling a comparison to be made between the two types of ozonation. Static mixers are fitted at both pre- and

Figure 4.1 Photograph of static mixer at Chertsey, UK (vertical configuration)

inter-ozonation stages at Chertsey (two separate pre-ozonation systems for the different streams), Egham and Walton. A photograph of the static mixers used at Chertsey is shown in (Figure 4.1). A static mixer is also used at Clay Lane for the single stage ozonation at that site. Also, a side stream static mixer is used at Clay Lane so a comparison between the different types of static mixer designs was also possible (Figure 4.2). A schematic of this arrangement is shown in Figure B.1 of Appendix B. In service, a small quantity of incoming water (about 0.6% of the mainflow) is taken by the motive water pump in the sidestream. The high velocity of flow through the eductor/ejector generates a vaccum for the ozone eduction. The gas/liquid stream then passes through the small static mixer in the side stream (about 100 mm in diameter) which operates at very high velocities and shear rate to form tiny gas bubbles (≅ 570 microns). These are then re-introduced into the main static mixer contactor which is designed to operate at an extremely low headloss (less than 0.3 m) and ensure intimate contacting between the gas and liquid phases. The main static mixer provides a "plug flow" environment and contact time for effective mass transfer of the gas into the liquid. This option provides for wide turndown (up to 6:1) as compared to "sole" static mixers in the main line (no sidestream) exhibiting high transfer efficiencies (> 99%).

Note: A schematic of the arrangement is shown in Figure B.1 of Appendix B and Table B.5 gives a cost comparison for a typical drinking WTP.

Figure 4.2 Photograph of static mixer at Clay Lane, UK (horizontal configuration-main static mixer, 1.0 m dia., headloss less than a foot. Sidestream has pump, eductor and a small static mixer.)

Details of the static mixers and ozone contactors are summarized in Table B.1 and Table B.2 shown in Appendix B. All of the systems utilize ozone generated from oxygen at a nominal concentration of 100g/Nm^3. The pre-ozone systems were designed for pre-oxidation rather than disinfection. At the pre-ozonation stage for the Chertsey Abbeymeads source the filter inlet channel is used for pre-oxidation. All the static mixers are gas water mixers (no sidestream) and are 316 L stainless steel, with elements 316Ti.

The sampling was categorized as "full" or "partial" sampling. During each four weekly period, each plant was sampled in one of these categories. Partial sampling took place at the inlet and the outlet of each of the ozonation process at each site. The following parameters were measured: turbidity, pH, UV absorbance (at 254 nm), alkalinity, TOC, total coliforms, aerobic spore forming bacteria, *Bacillus subtilis*, temperature, particle counts.

Full sampling took place at one of the sites in each of the four week periods. This involved sampling at each of the points carried out in the partial sampling and at up to three points within the ozone contactor (where possible within each of the chambers in the contact tank). In addition to the parameters listed previously, the following were also measured: bromide, bromate, atrazine, simazine, and total pesticides. The partial sampling generated a large quantity of data to show the effectiveness of the ozonation process under different inlet water conditions at each site. A total of 20 sample runs were completed at each site. Full sampling provided detailed information about the evolution of various parameters throughout the contactors. This was done four times at each site during the sampling program, so that the effect of differing conditions can again be seen. Samples were collected before and after each ozonation stage at each site and analyzed for the above parameters. In particular, intensive sampling was carried out in conjunction with the pilot-plant testing at the Chertsey WTP. All the analyses were carried out at VWP Central Laboratory, Watford. The data was used in the development and verification of the modeling aspects of the study.

Water Quality

The raw water for all the works except Clay Lane is drawn from different points along the River Thames (Chertsey also takes a proportion of its supply from Abbeymeads, a river gravel source). The raw water quality at the sites where data was collected is shown in Table 4.1. Atrazine concentration in the raw water ranged 0.01-0.21 μg/L at the WTPs. More details on the raw water quality can be found in Table B.3 of Appendix B.

Data Analysis

Spearman Rank Correlation

The Spearman rank correlation method is used to investigate the relationship between two independent variables that cannot be measured in a linear fashion. The calculation is based on the ranking of a variable, not on the variable itself. The calculated coefficient lies between ±1. For a perfect positive correlation, the coefficient equals 1, for a perfect negative correlation the coefficient equals –1. The closer the coefficient falls to zero the weaker the correlation.

The coefficient is defined as:

$$r = 1 - \frac{6\sum D^2}{n(n^2 - 1)} \qquad (4.1)$$

where

r = Spearman Coefficient
D = Difference in rank of the two variables
n = Number of data points

Table 4.1
Raw water quality at the WTP's

	Chertsey	Clay Lane	Egham	Iver	Walton
Alkalinity (mg/l as $CaCO_3$)	234-307	327-337	256-289	212-305	176-280
Bromide (μg/l)	115-281	90-209	75-164	73-133	66-157
pH	7.5	6.8-7.3	7.4-7.9	7.7-8.7	7.9-9.1
Temp (°C)	8-9	9-11	4-22	11-19	11-18
TOC (mg/l as C)	2.5-3.7	1.3-4.1	2.7-8.3	3.4-9.3	4.0-6.4
Turbidity (NTU)	0.1-0.6	0.1-2.4	0.3-115	0.2-64	1.0-78.1

RESULTS AND DISCUSSION

Effect of Ozone on pH, Alkalinity, and TOC

The results indicated that the ozonation process had no effect on pH and alkalinity. There were slight changes in the TOC value across the ozonation process. This is due to the effect of ozone chemically transforming the organic matter in the water, breaking high molecular weight compounds to low molecular weight compounds. This was expected as the plants use ozone doses of < 0.5 mg O_3/mg C.

Impact of Turbidity on Ozonation

Spearman rank analysis was performed on all the data collected to ascertain what the correlation was between the inlet turbidity and the ability of the ozonation process to inactivate or remove various parameters. The calculated coefficients indicated that there does not appear to be any correlation between inlet turbidity and the log inactivation of aerobic or *Bacillus* spores, UV254 or the removal of turbidity. There was a stronger negative correlation between inlet turbidity and particle count log removal. However, at high particle count values the upper limit of detection of the instrument was being approached and this negative correlation may be an artifact of the particle counter. Therefore, this data does not indicate that inlet turbidity had any effect on the ozonation process.

There was a distinct difference between the ability of the ozonation process to decrease the level of turbidity, depending on the quality of the inlet water. The pre-ozonation step with raw surface waters had little effect on the turbidity. However, this was not the case with the intermediate ozonation steps, or the pre-ozonation steps with groundwater. Here, the ozonation process resulted in a decrease in the turbidity of the sample. A similar trend was also seen with particle counts and the aerobic and *Bacillus* spores, which were removed much more easily from clean water than from more turbid water.

Inactivation of Total Coliforms, Aerobic Spores and **Bacillus subtilis** *by Ozone*

The log inactivation of total coliforms, aerobic spores and *Bacillus subtilis* across the ozone processes are shown in Tables 4.2 and 4.3. Total coliforms were removed entirely during the inter-ozone stage at Walton, Chertsey, Egham and Clay Lane. Unlike aerobic and *Bacillus*

Table 4.2
Log inactivation of total coliforms, aerobic spores and *Bacillus subtilis* across the pre-ozone process

Site	Walton	Chertsey N. Burway	Chertsey Abbey meads	Egham	Iver
Total Coliforms	1.29	1.14	N.A.	1.58	0.62
Aerobic spores	0.14	0.17	0.04	0.1	0.11
Bacillus subtilis	0.17	0.44	0	0.18	0.21

Table 4.3
Log inactivation of total coliforms, aerobic spores and *Bacillus subtilis* across the inter-ozone process

Site	Walton	Chertsey	Egham	Iver	Clay Lane
Total Coliforms	N.A.*	N.A.*	N.A.*	1.88	N.A.*
Aerobic spores	0.39	0.16	0.2	0.27	0.55
Bacillus subtilis	2.20	0.16	0.06	0.44	0.67

* NA=Not Applicable

spores, coliforms are removed instantaneously following the static mixer. The typical inactivation of coliforms through the ozone contactors is shown in Figure 4.3.

Only at Iver, coliforms were still present following inter-ozonation. At Iver, the water was more turbid while entering the inter-ozone stage than the other sites and this was the only site that used sparged mixing instead of a static mixer. However, even the log inactivation of coliforms across the pre-ozone step was significantly lower at Iver than at Walton, Chertsey (North Burway) and Egham. At Iver this inactivation was 0.6 log, whilst at the other sites it was 1.2, 1.1 and 1.6 log respectively. Since the quality of the water entering the pre-ozone contactors at Egham and Iver was similar, this indicates that the lower inactivation at Iver was not because of the quality of the incoming water. Indeed, as already discussed, Spearman rank analysis indicates that inlet turbidity did not affect the inactivation of biological parameters such as aerobic spores. The lower inactivation can be attributed to the conventional diffuser used instead of a static mixer.

The effect of ozone on microbial inactivation and bromate formation at the different sites are presented in Figures 4.4 to 4.7. The results for Chertsey WTP were presented in the previous Chapter. Due to the low concentration of spores in the raw water at these sites, it was deemed undesirable to compare the inactivation data as a function of Ct.

By looking at the reduction of spores through the inter-ozone stage, the need for an adequate residence time was demonstrated. Much of the spore inactivation takes place in the final compartment of the contact. This means that even with lower ozone doses (0.2 to 0.4 mg O_3/mg C), spore inactivation takes place, but at the limits of the residence time allowed by the contact tank. Inactivation of aerobic spores and *Bacillus subtilis* (0.10 to 2.20 Log) was observed throughout the ozonation processes during the overall sampling period at the different WTPs. Generally, there was a greater inactivation across the inter-ozone stage than the pre-ozone stage and greater inactivation of *Bacillus subtilis* than aerobic spores. It is important to note that, the

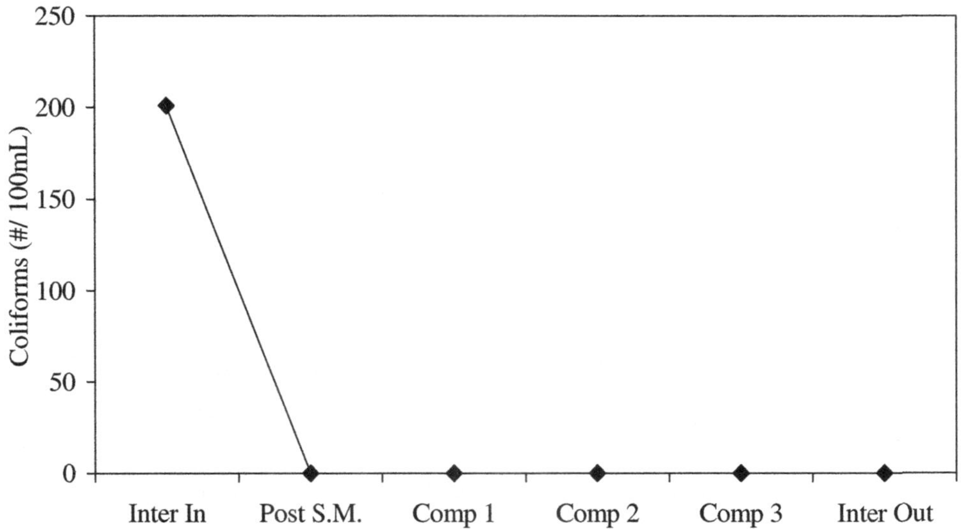

Figure 4.3 Coliform inactivation across inter-ozone stage at Egham (June 13, 2000)

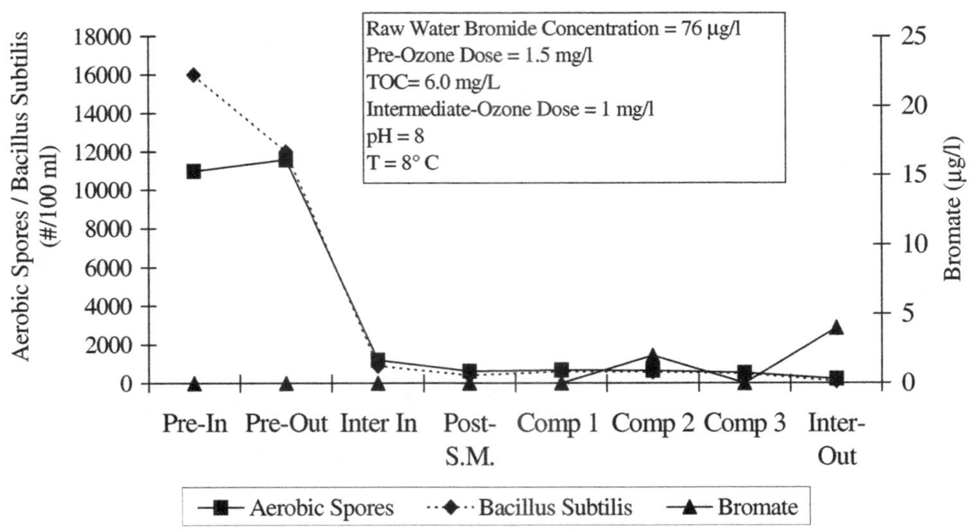

Figure 4.4 Effect of ozone treatment on aerobic spores, *Bacillus subtilis*, and bromate formation along Walton WTP (August 1999)

number of spores in the raw water was low 70 to 5000/100 mL. Only in some instances, they were as high as 10,000 to 48,000/100 mL. In these plants, ozonation was adopted mainly for pesticide destruction rather than for inactivation of spores, cysts or oocysts. The contact time in these plants varies between 1.0 and 13.1 minutes. Therefore, the residence time within the contactors was much shorter than would be necessary to effect complete inactivation of these biological spores. Tables 4.2 and 4.3 show the average microbial inactivation achieved at the pre- and intermediate-ozonation steps.

The influence of different parameters on aerobic spore inactivation across the inter-ozone contactors was assessed by Spearman rank analysis; the calculated coefficients are shown in

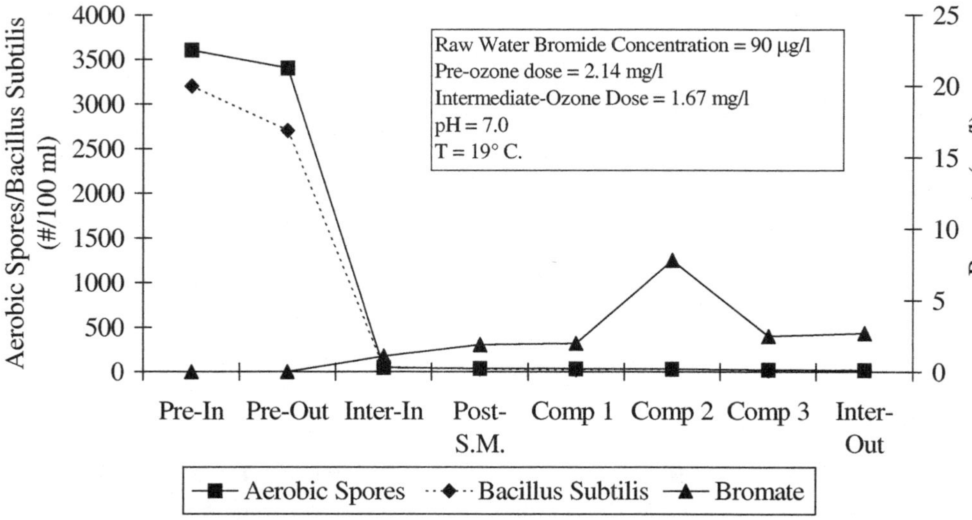

Figure 4.5 Effect of ozone on aerobic spores, *Bacillus subtilis*, and bromate formation along Egham Plant (September 1999)

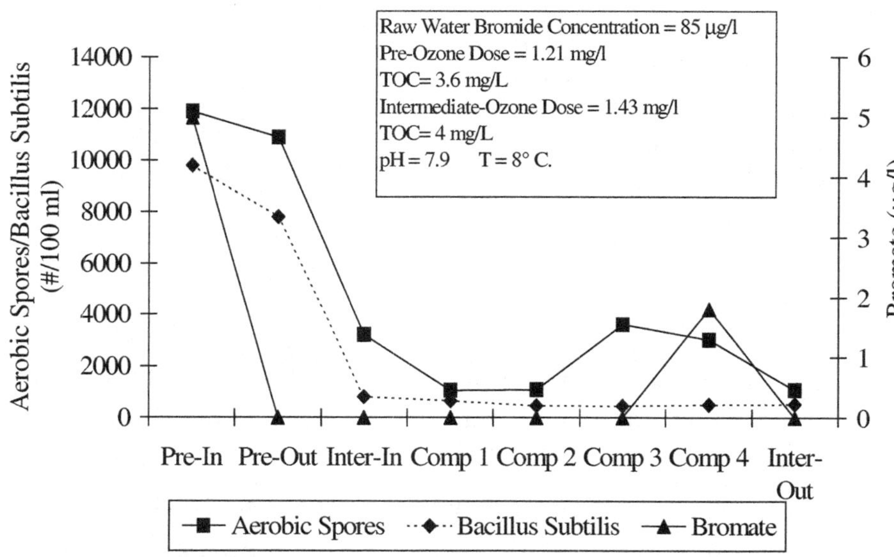

Figure 4.6 Effect of ozone on aerobic spores, *Bacillus subtilis*, and bromate formation along Iver Plant (February 2000)

Table 4.4. The strongest correlation was seen between ozone residual and spore inactivation. Thus, this data demonstrates that for adequate spore inactivation, it is important to have adequate residence time within the contactors and a sufficiently high ozone residual.

Formation of Bromate by Ozonation

Regulations governing bromate concentrations in drinking water in the UK are to be set at 10 µg/L. Therefore it is essential that the ozonation process be carefully controlled to ensure this limit is not exceeded. Bromate formation was depicted in Figures 4.4 to 4.7.

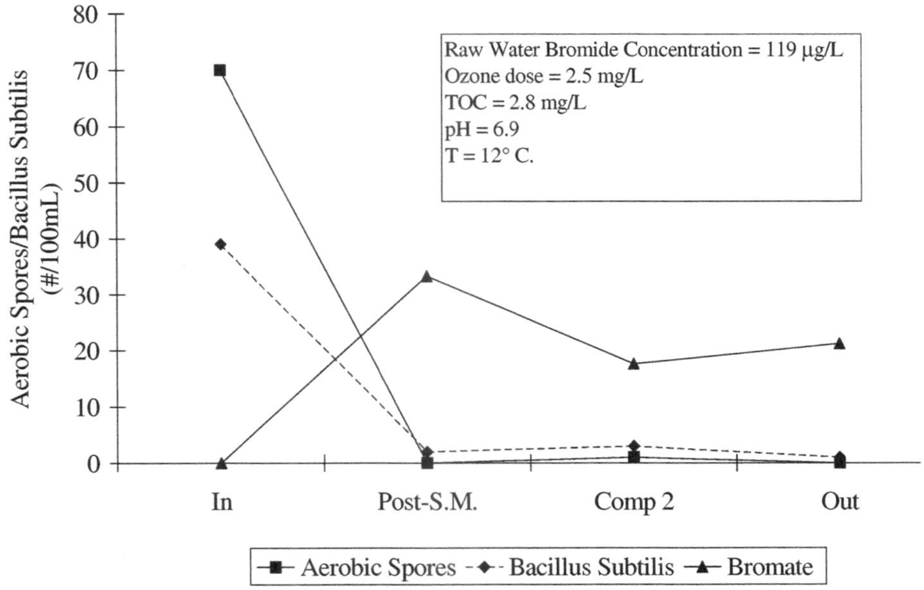

Figure 4.7 Effect of ozone on aerobic spores, *Bacillus subtilis*, and bromate formation along Clay Lane Plant (November 1999)

Table 4.4
Spearman rank coefficients for log inactivation of aerobic spores as a function of inlet turbidity, ozone dose and residual and inlet TOC level

Parameter	Spearman Rank Coefficient
Inlet Turbidity	-0.06
Ozone Dose	0.15
Ozone Residual	0.44
Inlet TOC	-0.01

At Chertsey, Egham and Iver the amount of bromate created was always below 3.5 µg/L, which is well below the proposed limit. At Walton, this limit was approached on occasion. The ozone dose for these plants varies between 0.2 and 0.4 mg O_3/mg C except for Clay Lane. This optimal dose ensured bromate below the MCL at all times even though the pH varied between 7.2 and 8.3. The Clay Lane plant uses a much higher ozone dose of 4 mg/L (0.7 to 1.4 mg O_3/mg C, typically controlling to a residual of 2 to 3 mg/L as opposed to around 0.4 mg/L at the other sites) which leads to bromate exceeding the MCL even at neutral pH. When the ozone dose was reduced to 0.4 mg/L, to combat the problem of bromate formation, the level of bromate exiting the ozone process was substantially decreased. The bromate levels through the Clay Lane contactors at these two ozone doses are shown in Figure 4.8. As previously discussed a longer residence time in the contact tank would accomplish greater pesticide removal and spore inactivation. However, Figure 4.8 clearly indicates that at higher ozone doses, bromate was formed throughout the contact time and the amount of bromate created increases with contact time. Therefore, an operational balance between the removal of pesticides, the inactivation of spores and the formation of bromate must be achieved.

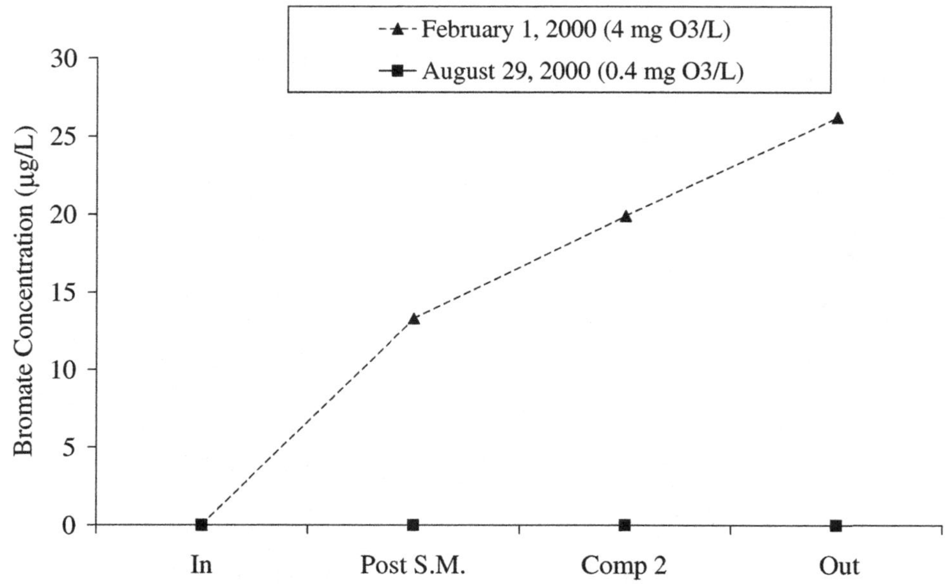

Figure 4.8 Bromate formation across inter-ozone stage at Clay Lane

Table 4.5
Spearman rank coefficients for amount of bromate formed as a function of ozone dose, ozone residual, inlet TOC and bromide levels

Parameter	Spearman Rank Coefficient
Ozone Dose (mg O_3 / L)	0.49
Ozone Dose (mg O_3 / mg C)	0.60
Ozone Residual	0.65
Inlet TOC	-0.02
Inlet Bromide	0.56

Spearman rank analysis was performed to determine the effect of ozone dose, residual, levels of TOC and bromide in the inlet water on bromate formation. The results are shown in Table 4.5. With the exception of TOC levels, there was a strong correlation between all the parameters examined and the amount of bromate formed. This reinforces the fact that the amount of ozone applied should be carefully controlled so as to minimize bromate formation.

Removal of Pesticides by Ozonation

Within the UK, the intermediate ozone process has been installed principally for pesticide removal. Urons are easily broken down by ozone, while triazines are more difficult to remove. Figure 4.9 illustrates the typical pesticide removal achieved through the intermediate ozone stage at Walton, Chertsey, Egham and Iver. "Total pesticides" is the cumulative value for all expected pesticides in the water. For the most part in these studies, it equates to the summation of the triazine pesticides, atrazine and simazine.

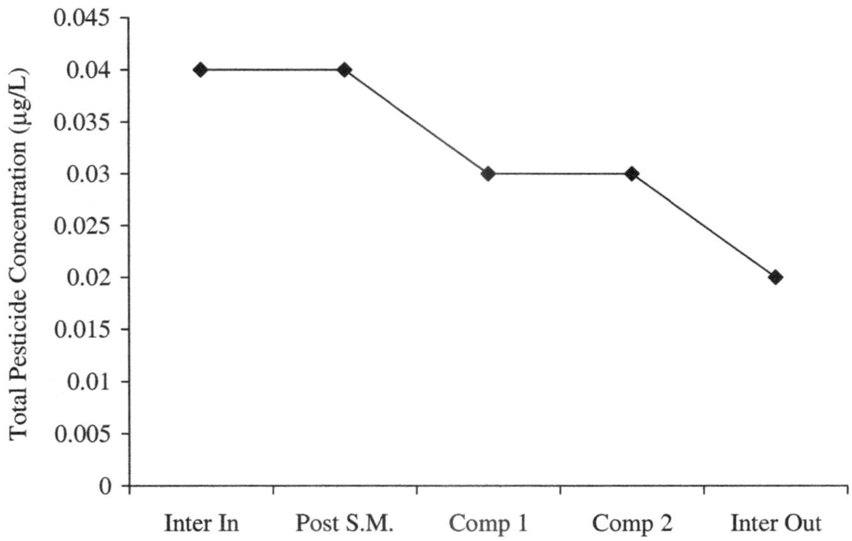

Figure 4.9 Total pesticide removal across inter-ozone stage at Walton (October 11, 2000)

Like aerobic spores and *Bacillus subtilis*, there is a continuous removal of total pesticides throughout the contactors. This suggests that if there were a greater residence time within the contact tank, pesticides would be completely removed. At all the sites pesticide removal is a two stage process involving ozone followed by GAC contactors At Walton, Chertsey, Egham and Iver the amount of total pesticides in the water following intermediate ozone was always below 0.05 µg/L.

At Clay Lane the pesticide levels in the inlet water are at least three times higher than at the other sites and the effectiveness of the ozone process for pesticide removal is more critical. Data collected as part of another study indicates that the ozone process at Clay Lane removes urons. However, as Figure 4.10 indicates, it does not remove triazines as effectively. The reason for the zero results immediately following the static mixer have not been ascertained. It is likely that these results are an artifact of the sampling procedure and not a genuine result. There was only an average removal for atrazine and simazine at Clay Lane of 25% and 27% respectively when a high dose of 4.0 mg/L was applied.

The intermediate ozonation treatment in Walton enabled a removal of 50% for total pesticides and a complete destruction of simazine (Figure 4.11). Complete inactivation of *Bacillus subtilis* was ensured at the outlet of the intermediate ozonation stage while keeping the bromate formation under the MCL at an ozone dose of 0.3 mg O_3/mgC.

Spearman rank coefficients to correlate pesticide removal with various parameters indicated that there is a slight correlation between ozone dose and pesticide removal, although this is not as high as might be expected.

SUMMARY

Ozonation had no effect on pH or alkalinity but slightly reduced the incoming TOC. Static mixers were found to be more effective for biological inactivation than conventional diffusers, particularly for the inactivation of coliforms. Looking at the log inactivations of coliforms across

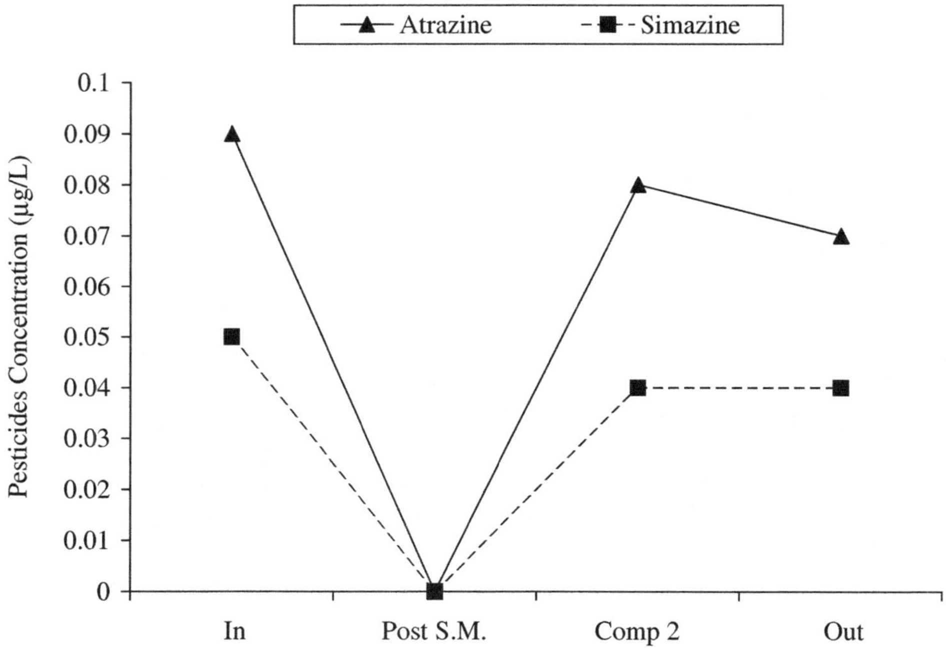

Figure 4.10 Triazine removal across the ozone contactors at Clay Lane (February 1, 2000)

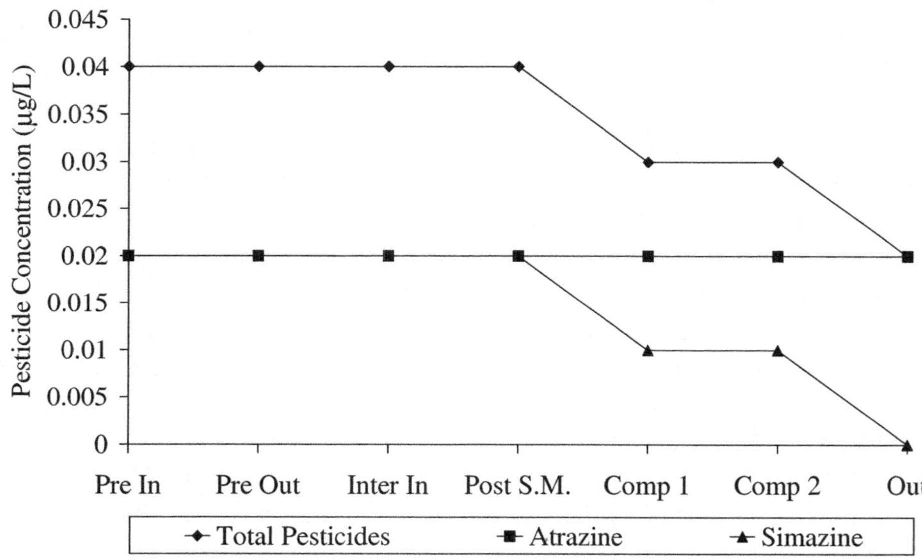

Figure 4.11 Pesticide oxidation through pre- and inter-ozonation at the Walton WTP (October 2000)

the pre-ozonation steps at the sites where this is applicable, it is clear that the inactivation by the Iver ozonation process is significantly less than that at Egham, Chertsey and Walton. Also, the least inactivation of aerobic spores and *Bacillus subtilis* was observed at this plant. Iver is the only one of the sites under consideration that utilizes turbine diffusers and porous plate diffusers for ozone dissolution instead of static mixers and shows coliforms at the outlet of the intermediate ozone stage. This result is not a result of the quality of the incoming water, which is similar at Iver and Egham. Indeed, the higher individual log inactivations for coliforms at Iver come at times of higher turbidity.

Inactivation of coliforms across the inter-ozone stages was almost instantaneous. Spores and pesticides were inactivated or reduced throughout the contact tank, indicating that inactivation and removal of these parameters by ozone is influenced by the residence time in the contact tank. On the whole, the log removal of particles was of the same order as the log inactivation of the spores across the ozonation processes. Both particle counts and aerobic spores have been used as surrogates for *Cryptosporidium* oocysts in the past. Therefore, the fact that they behaved similarly through the treatment process was encouraging.

The results for the inactivation of aerobic and *Bacillus subtilis* spores varied between 0.2 and 2.2 log at these sites. In general, *Bacillus subtilis* was less resistant to ozone than the aerobic spores. In these plants, ozonation was adopted mainly for pesticide destruction rather than for inactivation of spores, cysts or oocysts. The contact time in these plants varies between 1.0-13.1 minutes. Therefore, the residence time within the contactors is much shorter than would be necessary to effect complete inactivation of these biological spores. The ozone dose for these plants varied between 0.2 and 0.4 mg O_3/mg C except for Clay Lane. An optimal ozone dose of 0.2 to 0.4 mg O_3/mg C ensured bromate below the MCL at all times even though the pH varied between 7.2 and 8.3. The Clay Lane plant uses a much higher ozone dose 0.7 to 1.4 mg O_3/mg C (typically controlling to a residual of 2 to 3 mg/L as opposed to around 0.4 mg/L at the other sites) which lead to bromate exceeding the MCL at neutral pH. Lately bromate formation has been reduced to less than the MCL at this plant by lowering the ozone dose (residual = 0.1 mg/L) with additional process improvements. Bromate formation was shown to be influenced by the contact time, ozone dose, pH and the amount of bromide in the inlet water.

Triazine pesticides were not removed by ozonation. However, by removing other material which would compete for sites within the GAC filters, pesticide removal was aided by the ozonation process.

CHAPTER 5
HYDRODYNAMIC, MASS TRANSFER AND PESTICIDE DESTRUCTION MODELING

INTRODUCTION

In 1998, the first model for static mixers was built by Langlais (1998) with a focus on the micro-pollutant removal efficiency (i.e. atrazine) of the hydrogen peroxide-ozone advanced oxidation process. The goal of the present study was to improve the model in order to estimate the ozone transfer and consumption through the static mixer and the gas-liquid separator. The model was based on a physico-chemical description of the various phenomena occurring in the contactor. To ensure a design capability, the required parameters were planned to be obtainable by other means (such as Residence Time Distributions, kinetic batch studies or geometrical analyses, *Cryptosporidium* inactivation) than an adjustment of the model to some laboratory measurements. Once the extension of the model was completed, the validation was planned to cover a large variety of laboratory situations.

MODEL DESCRIPTION

Two components have to be combined to build a comprehensive model of static mixer contactor systems: the chemistry of ozone in natural water and the physics of the two-phase flow through the static mixers as well as in the gas-liquid separator.

Chemistry of Ozone

Basic Ozone Chemistry in Water

The action of ozone is usually divided into a molecular, or direct, reaction pathway where the ozone molecule interacts directly with some selectively oxidizible compounds, and a indirect reaction pathway where ozone decomposes into hydroxyl radicals which in turn react with a large variety of organic and inorganic compounds.

The degradation reaction between molecular ozone or hydroxyl radicals with a given compound, C, usually is a global second order reaction, which can be described by the following kinetic expressions:

$$-k_{D,C} \cdot [O_3] \cdot [C]$$

$$-k_{ID,C} \cdot [OH^\circ] \cdot [C]$$

However, the transformation of molecular ozone into hydroxyl radicals obeys a much more complex mechanism (Hoigné and Bader, 1983a,b): a radical chain reaction initiated by several chemical species (hydroxide ions, hydrogen peroxide, ferrous ions, formiate ions and various organic functional groups) takes place. Among those compounds, two are naturally occurring and unavoidable: the hydroxide ions (depending on pH) and the natural organic matter. Hydrogen peroxide might also be added for an advanced oxidation process.

In pure water, several ozone decomposition mechanisms have been suggested (Stahelin et al., 1984; Gordon 1987) with an overall similar behavior. We will be using the one developed by Stahelin and Hoigné (1982):

- Non radical reactions:

$H_2O \Leftrightarrow HO^- + H^+$ $\quad pK_{A1} = 14$

$H_2O_2 \Leftrightarrow HO_2^- + H^+$ $\quad pK_{A2} = 11.6$

$H_2CO_3 \Leftrightarrow HCO_3^- + H^+$ $\quad pK_{A3} = 6.3$

$HCO_3^- \Leftrightarrow CO_3^{2-} + H^+$ $\quad pK_{A4} = 10.25$

- Initiating reactions:

$O_3 + OH^- \rightarrow O_2^{-\bullet} + HO_2^\bullet$ $\quad k_1 = 70 \pm 0.7\ M^{-1}s^{-1}$

$O_3 + HO_2^- \rightarrow O_3^{-\bullet} + HO_2^\bullet$ $\quad k_2 = (2.8 \pm 0.5) \cdot 10^6\ M^{-1}s^{-1}$

- Propagation reactions:

$O_3 + O_2^{-\bullet} \rightarrow O_3^{-\bullet} + O_2$ $\quad k_3 = (1.6 \pm 0.2) \cdot 10^9\ M^{-1}s^{-1}$

$O_3 + HO^\bullet \rightarrow HO_4^\bullet$ $\quad k_4 = (2.0 \pm 0.5) \cdot 10^9\ M^{-1}s^{-1}$

$H_2O_2 + HO^\bullet \rightarrow HO_2^\bullet + H_2O$ $\quad k_5 = (2.7 \pm 0.2) \cdot 10^7\ M^{-1}s^{-1}$

$HO_2^- + HO^\bullet \rightarrow HO_2^\bullet + HO^-$ $\quad k_6 = (7.5 \pm 0.3) \cdot 10^9\ M^{-1}s^{-1}$

$HO_4^\bullet \rightarrow HO_2^\bullet + O_2$ $\quad k_7 = (2.8 \pm 0.3) \cdot 10^4\ M^{-1}s^{-1}$

$HO_2^\bullet \Leftrightarrow O_2^{-\bullet} + H^+$ $\quad pK_{A5} = 4.8$

$O_3^{-\bullet} + H^+ \rightarrow HO_3^\bullet$ $\quad k_8 = (5.2 \pm 0.6) \cdot 10^{10}\ M^{-1}s^{-1}$

$HO_3^\bullet \rightarrow O_3^{-\bullet} + H^+$ $\quad k_9 = (3.3 \pm 0.3) \cdot 10^2\ s^{-1}$

$HO_3^\bullet \rightarrow HO^\bullet + O_2$ $\quad k_{10} = (1.1 \pm 0.1) \cdot 10^5\ s^{-1}$

$HCO_3^- + HO^\bullet \rightarrow HCO_3^\bullet + HO^-$ $\quad k_{11} = (1.5 \pm 0.2) \cdot 10^7\ M^{-1}s^{-1}$

$CO_3^{2-} + HO^\bullet \rightarrow CO_3^{-\bullet} + HO^-$ $\quad k_{12} = [2\ ;\ 4.8] \cdot 10^8\ M^{-1}s^{-1}$

$HCO_3^\bullet \Leftrightarrow CO_3^{-\bullet} + H^+$ $\quad pK_{A6} = 9.6$

- Scavenging reactions:

$2\ CO_3^{-\bullet} \rightarrow CO_4^{2-} + CO_2$ $\quad k_{13} = 1 \cdot 10^7\ M^{-1}s^{-1}$

$HO^\bullet + HO_2^\bullet \rightarrow H_2O + O_2$ $\quad k_{14} = 1.1 \cdot 10^{10}\ M^{-1}s^{-1}$

Considering the very short lifetime of the radical species, Orta de Velasquez (1992) used a quasi-stationary assumption to develop an algebraic relation between the hydroxyl radical concentration and the dissolved ozone concentration:

$$\left[HO^{\bullet}\right]=\frac{2\cdot k_1\cdot 10^{pH-14}+2\cdot k_2\cdot 10^{pH-11.6}\cdot\left[H_2O_2\right]}{(k_{11}+k_{12}\cdot 10^{pH-10.25})\cdot[HCO_3^-]}\cdot\left[O_3\right] \tag{5.1}$$

Reaction With the Natural Organic Matter

In addition to this basic set of reactions, the natural organic matter also plays an important role. It can be divided into subgroups: M_i the unknown organic compounds reacting directly with ozone, I_i the unknown organic compounds initiating the radical mechanism, P_i the unknown organic compounds promoting the radical mechanism, and S_i the unknown organic compounds scavenging the hydroxyl radicals. The additional set of reactions is therefore:

$M_i + O_3 \rightarrow M_{i,ox} + O_2$	k_{D,M_i}
$I_i + O_3 \rightarrow I_{i,ox} + HO^{\bullet}$	k_{D,I_i}
$P_i + HO^{\bullet} \rightarrow P^{\bullet}_{i,ox} \rightarrow P_{i,ox} + HO_2^{\bullet}$	k_{ID,P_i}
$S_i + HO^{\bullet} \rightarrow S_{i,ox}$	k_{ID,S_i}

Of course, compounds can belong to several groups simultaneously. For some well-known compounds, the reaction rate constants have been determined, mostly small simple molecules and pesticides.

The ozone - hydroxyl radical correlation is modified accordingly:

$$\left[HO^{\bullet}\right]=\frac{2\cdot k_1\cdot 10^{pH-14}+2\cdot k_2\cdot 10^{pH-11.6}\cdot[H_2O_2]+\sum_i k_{D,I_i}\cdot[I_i]}{(k_{11}+k_{12}\cdot 10^{pH-10.25})\cdot[HCO_3^-]+\sum_i k_{ID,S_i}\cdot[S_i]}\cdot\left[O_3\right] \tag{5.2}$$

A difference has been made between a long-term ozone demand and a short term or "instantaneous" ozone demand. The ability to determine the ozone decomposition rate constants by the ozone profile in batch experiments is, more or less, the arbitrary segmentation criteria.

The long term ozone demand kinetic induced by the natural water has been globalized into a first order reaction rate:

$$r_{O_3} = -w\cdot\left[O_3\right] \tag{5.3}$$

The w apparent kinetic constant contains the various long-term contributions of the water:

- the direct and indirect ozone consumption by the naturally occurring organic matter;
- the initiating effect of the pH; and
- the scavenging effect of the carbonates.

Table 5.1 presents the various relations that have been developed based on laboratory studies.

The short term ozone demand can be deduced from batch experiments: Roustan et al. (1997) found values covering the range between 0 and 1.5 mg/L depending on the water source and the treatments preceding the ozonation. Duguet (2000) studied the instantaneous ozone demand of Seine River water and showed that usual treatments, like clarification and filtration, reduced the demand.

The kinetics of the so called "instantaneous" ozone demand is usually neglected in gas/liquid separators: the satisfaction of that demand takes at the most 20 seconds compared to the column residence time of several minutes. However, in static mixers the two time frames overlap and have to be considered. Muñoz Ramirez (1997) obtained kinetic rate constants through competitive ozonation of the natural water and of a measurable reference compound, whose degradation rate constant was known. Several waters, from the Paris suburb and from the Rennes area, were studied. A first order reaction rate was assumed:

$$r_{O_3} = -\kappa \cdot [O_3] \quad (5.4)$$

Table 5.1
Laboratory correlations for the ozone decay kinetic constant

$\log w\ (h^{-1}) = -3.98 + 0.66\ pH + 0.61 \log(TOC) - 0.42 \log(alkalinity/10)$				
Laboratory conditions				Yurteri and Gurol, 1988
Synthetic solutions with fulvic acids				
Magnetically stirred discontinuous reactor				
T=20±1°C	6.8 <pH< 9	0.3 < TOC < 5 (mg/L)	10< Alkalinity <500 (mg $CaCO_3$/L)	
$\log w(s^{-1}) = -3.93 + 0.24\ pH + 0.7537 \log(UV_{254}) + 1.08 \log(TOC) - 0.19 \log(alkalinity)$				
Laboratory conditions				Orta De Velasquez, 1992
Natural waters				
Magnetically and mechanically stirred discontinuous reactor				
T=15°C	7 <pH< 8.7 after treatment	2 < TOC < 5.5 (mg/L)	50 <Alkalinity < 250 (mg $CaCO_3$/L)	0.05 <DO_{254}<0.4
$\log w\ (min^{-1}) = -4.0 + 0.29\ pH + 1.19 \log(TOC) - 0.41 \log(alkalinity)$				
Laboratory conditions				Wang, 1995a
Tap water				
Gas/liquid separator				
T=15°C	7.5 <pH< 8.1	2 < TOC < 6 (mg/L)	84 <Alkalinity< 150 (mg $CaCO_3$/L)	

The apparent rate constant κ depends on the nature of the natural organic matter, thus its value will be a function of the water's origin, period of the year, upstream treatments, applied ozone dose and pH. The observed initial values for κ were in the range 0.5 to 3 s^{-1}, while the long-term reaction rate constants w are all below 0.02 s^{-1}.

Like for the long-term ozone demand, it is expected that some of the organic matter does react as radical initiator. At this time, we have or we know how to obtain values for:

- the long term ozone demand (Yurteri and Gurol, 1988; Orta de Velazquez, 1992; Wang, 1995a);
- the hydroxyl radicals initiated by the long term demand (Levilly, 1999; Savary et al., 2000);
- the amount of "instantaneous" demand (Roustan et al.(1997) , Duguet (2000)); and
- the kinetics of the fast ozone decay (Muñoz Ramirez , 1997).

Hence, if we assume that the "instantaneous" radical production is the only missing phenomena, we will be able to deduce its value using a tracer compound, whose reaction rate constants with ozone and hydroxyl radicals is well determined. The water to be treated can be spiked with such a compound. Once the various known contributions to its degradation estimated, the excess consumption observed in the laboratory can be attributed to the "instantaneous" radical contents. If β is the fraction of "instantaneous" demand leading to hydroxyl radicals, the relation between ozone and radicals becomes:

$$\left[HO^{\bullet}\right]=\frac{2\cdot k_1\cdot 10^{pH-14}+2\cdot k_2\cdot 10^{pH-11.6}\cdot[H_2O_2]+K_D\cdot[TOC]+\beta\cdot\kappa}{(k_{11}+k_{12}\cdot 10^{pH-10.25})\cdot[HCO_3^-]+K_{ID}\cdot[TOC]}\cdot[O_3] \qquad (5.5)$$

Mass Transfer

When a reaction needs the transfer of a reactant through an interface, the efficiency of that reaction depends simultaneously on the intrinsic reaction and on the mass transfer. In contactors with a low mixing intensity and fast reactions, the transfer rate is often slower than the reaction kinetics. It becomes the limiting factor. However, in static mixers, because of the high mixing intensity, both factors are significant.

The mass transfer between two fluid phases has been modeled by numerous authors based on various assumptions (Coulson and Richardson, 1993). Some of the most used models are described below:

- The double layer theory (Whitman, 1923; Lewis and Whitman, 1924) assumes that the transfer resistance is localized in two layers, stagnant on each side of the interface and of thickness δl and δg. In each of these layers, the transport is exclusively ensured by molecular diffusion.
- The penetration theory of Higbie (1935) assumes a renewal of small size fluid elements from the interface by new ones from the fluid bulk. Danckwerts (1951) modified this model introducing a residence time distribution of the fluid elements: the surface renewal theory.

- The film-penetration theory of Toor and Marchello (1958) combines both approaches.
- Other models exist like those proposed by Hishinevskij (1951) or Wang and Langemann (1994a).

All these models have been adapted for physical absorption with chemical reactions (Hatta, 1932; Danckwerts, 1955; Huang and Kuo, 1963; Wang and Langeman, 1994b).

The Double Diffusional Layer Theory

In the double layer theory, the transfer resistance is localized in the boundary layers, on each side of an interface between two fluids. While mixing in the bulk of the fluid phase is mainly due to convective transport and turbulent diffusion, within these layers only molecular diffusion occurs. The law governing the molecular diffusion is the Fick's law:

$$N_A = -D_{A,l} \frac{d[A]}{dz} \tag{5.6}$$

This relation expresses the transfer of a component A in some media l along the z direction when transport can be neglected and with the assumption that molecular opposite diffusion occurs. Under stationary condition, and introducing a linear approximation for the concentration profile, Fick's Law can be integrated as follows:

$$N_A = \frac{D_{A,g}}{\delta_g . RT}\left(P_A - P_A^*\right) = \frac{D_{A,l}}{\delta_l}\left([A]^* - [A]\right) \tag{5.7}$$

Both expressions are necessarily equal for continuity reasons. P_A* and [A]* are the pressure and the concentration at the interface and are related by the gas-liquid equilibrium. P_A and [A] are the bulk pressure and concentration.

The mass transfer coefficients, k_g and k_l, can then be defined for each side:

$$k_g = \frac{D_{A,g}}{\delta_g} \quad \text{and} \quad k_l = \frac{D_{A,l}}{\delta_l}$$

An increase in turbulence within the bulk reduces the boundary layer thickness and, the transfer coefficient accordingly.

For compounds with a low solubility such as ozone, the gas-liquid equilibrium can be estimated by the Henry's law. Overall transfer coefficient can then be defined by:

$$N_A = \frac{K_g}{RT}(P_A - P_{A,e}) = K_L\left([A]_e - [A]\right) \tag{5.8}$$

with $\dfrac{1}{K_g} = \dfrac{1}{k_g} + \dfrac{He'}{RT.k_l}$ and $\dfrac{1}{K_L} = \dfrac{1}{k_l} + \dfrac{RT}{He'.k_g}$

$P_{A,e} = He'.[A]$ and $[A]_e = P_A / He'$

When a compound easily diffuses into the gas phase and with difficulties in the liquid phase ($k_l << k_g$) like for ozone in water, then the overall resistance is located in the liquid boundary layer:

This simplifies the model to $P_A^* = P_A$, $[A]^* = [A]_e$ and $\frac{1}{K_L} = \frac{1}{k_l}$

In that case, the local volume transfer rate is given by:

$$dN_A = k_l \cdot a \cdot ([A]^* - [A]) = K_L \cdot a \cdot ([A]_e - [A]) \quad (5.9)$$

where a is the superficial area per volumetric unit.

Static Mixers

The transfer of ozone in water using static mixer as in-line contactor has been studied by several authors: Grozs-Röll et al., (1983); Martin, (1992); Jamin-Changeart, (1992); Hérès, (1993); Glover et al., (1993); Schullz and Prendiville, (1993); Langlais, (1998); Bonnard (1998). Numerous types of static mixers are available for industrial use: Sulzer, Kenics, Lightnin, Ross, Statiflo, Helax, Komax. The efficiency of these various modules can be quite different and does depend on the hydraulic diameter d_h. A small hydraulic diameter implies a large wetted surface, a large k_la and an increase of the power consumption (Bourne and Maire, 1991). The value of d_h of most helicoïdal mixers is related to the internal diameter of the pipe, i.e. $d_h \cong 0.5\ d_i$ (Rader et al., 1989). For mixers consisting of embossed sheets (i.e Sulzer), the hydraulic diameter is independent of the pipe diameter, and can be adjusted by modifying the gap between the sheets or the embossing.

In order to limit the head losses, the mixer elements are alternated with void spaces. This has little influence on the mass transfer coefficient k_l but does significantly affect the interfacial area. If l_S and l_V represent respectively the length of the static mixing element and of the void between two elements, the following schematic depicts the general configuration of the static mixer:

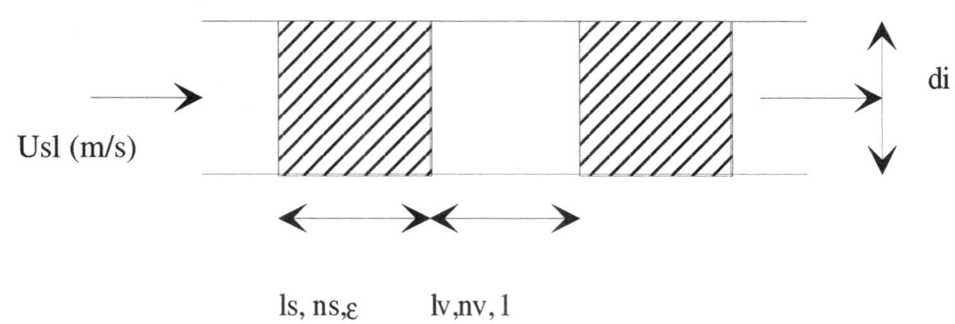

reactor's volume or dispersion volume: $V_d = (n_s.\varepsilon.l_s + n_v.l_v).A$
global length of the static mixer device: $l_{SM} = n_s.\varepsilon.l_s + n_v.l_v$
liquid superficial velocity: $U_{sl} = Q_L / A$

internal liquid velocity: $U_l = U_{sl}/\varepsilon_L$.

with: $A = \frac{\pi}{4}.d_i^2$

ε: void fraction of a mixing element

ε_L: fraction of volume of the pipe used by the liquid

Headloss

The headloss generated by the static mixers has to be overcome. Thus, headloss is key factor in the estimation of the efficiency: it does represent the energy dissipated for the transfer. Parameters affecting the head loss are (Zhu et al., 1989b):

- liquid flow rate;
- viscosity and density of the fluids;
- gas - liquid flow rate ratio; and
- shape, number and configuration of the mixing elements.

For single-phase flows, the headloss can be estimated by:

$$\Delta P = Ne_{SM}.\rho.l_{SM}.Us^2 \frac{l_{SM}}{di} \qquad (5.10)$$

$$\text{and} \quad Ne_{MS} = 2.f$$

with f, the friction factor, which has a constant value, for given elements and diameter, when the flow regime is turbulent, while, for laminar flow, it is proportional to the Reynolds number (Shah and Kale, 1992, Chandra and Kale, 1992).

$$Re = \frac{U_l.d_i}{\upsilon_l} = \frac{U_{Sl}}{\varepsilon}.\frac{\rho_l}{\mu_l}.d_i$$

$$Re_{d_h} = \frac{U_l.d_h}{\upsilon_l}$$

For Newtonian fluids in a turbulent flow, Martin (1992) obtained with superficial liquid velocities in the range 0.5 to 5 m.s^{-1}:

$f_{Sülzer} = 0.3$ ($D/d_h = 3$)

$f_{Kenics} = 0.4$

($f_{Kenics} = 0.49$, Studer, 1991)

Usually, f values between 0.4 and 0.5 are used.

For two phase flows in a horizontal pipe, the headloss is deduced from the head losses of each phase corrected by the Lockhart and Martinelli factor (Grosz-Röll et al., 1983):

$$X = \sqrt{\frac{\Delta P_l}{\Delta P_g}} \qquad (5.11)$$

Using the graphs on Figure 5.1 and X, the values of ϕ_l and ϕ_g are obtained and the overall headloss is determined:

$$\Delta P_{SM} = \phi_l^2.\Delta P_l = \phi_g^2.\Delta P_g\,. \qquad (5.12)$$

If the mixers are set up vertically, the hydraulic factor has to be added (upward flow) or subtracted (downward flow):

$$\Delta P = \Delta P_{SM} \pm \rho_l.G_{SM}(1-\varepsilon_g) \qquad (5.13)$$

Power Requirement

The required power is a comparison criteria among the various static mixer types. It can be estimated using (Hearn, 1993):

$$W = Q_L.\Delta P + Q_G\,\frac{2.P_{out}+\Delta P}{2}.\ln\left(\frac{P_{out}+\Delta P}{P_{out}}\right)+\frac{1}{2}.\rho_L\,Q_L.U_{s,1}^2.\left(1-\frac{1}{(1-\varepsilon_g)^2}\right) \qquad (5.14)$$

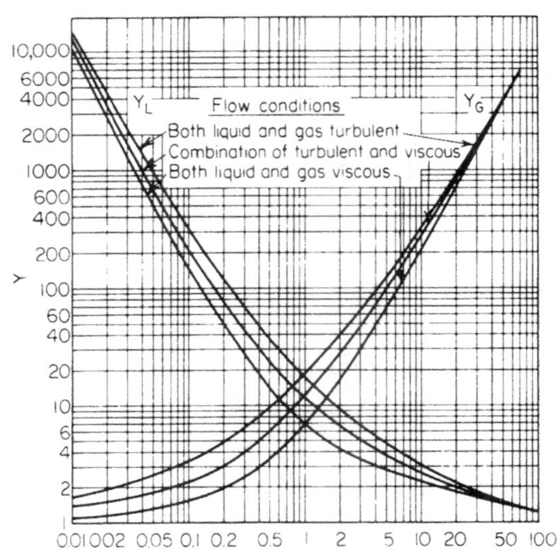

Figure 5.1 Abacus of Lockhart and Martinelli (1949)

Gas Holdup

The gas holdup εg, is defined as for the liquid by following set of equations:

$$\begin{cases} \varepsilon_g.U_g = U_{s,g} \\ \varepsilon_l.U_l = U_{s,l} \\ \varepsilon_g + \varepsilon_l = \varepsilon \approx 1 \end{cases}$$

If we include the slipping velocity, which expresses, for a vertical flow, the difference between the gas and the liquid velocity ($U_{gliss} = U_l - U_g$) due to their specific gravities. It is a function of the bubble shape resulting from several forces:

- inertial force;
- viscous force;
- gravity force; and
- surface tension.

A second degree equation is obtained whose positive solution is:

$$\varepsilon_g = \frac{\sqrt{(U_{s,l} + U_{s,g} - U_{gliss})^2 + 4.U_{s,g}.U_{gliss}} - (U_{s,l} + U_{s,g} - U_{gliss})}{2.U_{Gliss}} \quad (5.15)$$

For a horizontal flow the system can be considered homogeneous and $U_{gliss} = 0$ (Grosz-Röll et al., 1982). The gas holdup is then given by:

$$\varepsilon_g = \frac{U_{s,g}}{U_{s,l} + U_{s,g}} \approx \frac{Q_G}{Q_L + Q_G} \quad (5.16)$$

In general, the slipping velocity is fixed at ± 0.3 $m.s^{-1}$, however for small gas to liquid ratios, slipping can be neglected in static mixers (Roes et al., 1984).

Bubble Size and Interfacial Area

The bubble population can be represented by its Sauter diameter d_{32} defined by:

$$d_b \approx d_{32} = \sum_{i=1}^{p} d_i^3 \Big/ \sum_{i=1}^{p} d_i^2 \quad (5.17)$$

with d_i: individual bubble diameters
p: number of bubbles

In static mixers the bubbles are not necessarily smaller but their distribution is narrower, even for slow superficial velocities: 80 % of the bubbles have a diameter comprised between 0.7 d_b and 1.3 d_b. The bubble diameter for an air-water system and a dissipated energy of 10 W/kg is about 1 mm (Zhou et al., 1992). Table 5.2 gathers some correlations of the bubble diameter, notably in function of the Weber number:

$$We = \frac{\rho_L \cdot U_{sl}^2 \cdot d_h}{\varepsilon^2 \cdot \sigma} \tag{5.18}$$

The evaluation of the bubble diameter is difficult because of the approximations about the gaseous retention ε_g, the interfacial tension σ, and the mixer efficiency η. For guidance, this parameter is estimated at 0.15 for Sülzer SMV (Hérès, 1993) and at 0.17 for Kenics (Hearn, 1993).

The alternative use of elements and of void spaces has a great impact on the bubble size. The interfacial area variation has a sinusoidal like evolution between a_{max} and a_{min} (Turunen and Haario, 1994) and is not reproduced by the below mentioned correlation. The interfacial area and the bubble diameter are linked by Calderbank's relation (1967):

$$a = \frac{6.\varepsilon_g}{d_b} \tag{5.19}$$

Table 5.2
Correlations for the estimation of the mean bubble diameter in static mixers

Correlation	Conditions	References
$\frac{d_{32}}{d_h} = 0.21\, We^{-0.5}\, Re^{0.15}$	Sülzer SMV	Grosz-Röll et al., 1983
$\frac{d_{32}}{d_h} = 0.21\, We^{-0.5} Re^{0.15} (1+1.7\,\varepsilon_g)$	Sülzer SMR	Steiff et al, 1991
$\frac{d_{32}}{d_i} = C^{ste}.We^{-0.6}.Re^{0.1}$	Kenics Re>3.10^3	Zhou et al., 1992
$\frac{d_{32}}{d_h} \approx We^{-0.5}$	Lightnin	
$d_{32} = 0.4 \frac{\sigma^{0.6}}{(\rho_l^2.\rho_g^2)^{\frac{1}{2}}} \cdot \left[\eta . \frac{W}{\rho_l . V_d .(1-\varepsilon_g)}\right]^{-0.4} .(1+1.7\,\varepsilon_g)$	Sülzer	Roes et al., 1984 Russel et al., 1987 *(Cited by Hérès, 1993)*
$d_{32} = 0.4 \frac{\sigma^{0.6}}{(\rho_l^2.\rho_g^2)^{\frac{1}{2}}} \cdot \frac{W}{\rho_l . V_d .(1-\varepsilon_g)} .(1+0.4\,\varepsilon_g)$	Kenics coalescing system	Vernon, 1995
$d_{32(non\ coalescing)} = 0.65\ d_{32(coalescing)}$	non coalescing system	

Mass Transfer Coefficients

The product between the global mass transfer coefficient k_l and the interfacial area was the topic of many studies. Usually, the transport example used is the transfer of the oxygen contained in air into water. However some studies concerning the transfer of ozone into water are also available: Grosz-Röll et al., (1983); Jamin-Changeart, (1992); Hérès, (1993); Martin, (1992). Some of these correlations are shown in Table C.1 of Appendix C. The dissipated power is an important parameter for the estimation of the mass transfer coefficient because it is related to the turbulence within the reactor.

In a static mixer, the range for the k_la varies from 0.1 to 10 s^{-1}. Transfer coefficient of about 5 s^{-1} for 400 W/kg of dissipated power have been reported (Middleton, 1978), which is 100 to 1000 times greater than for a baffled reactor or a classical gas/liquid separator (Hérès, 1993).

Mass Transfer Acceleration by Fast Reactions

If a transferring compound A, initially in the gas phase, reacts with another product B in the liquid phase, the total transferable amount of A will increase, but in addition the mass transfer rate can also be enhanced. An acceleration factor E is defined by:

$$E = \frac{k_l(\text{with reaction})}{k_l(\text{without reaction})}$$

For the following reaction $\upsilon_A A + \upsilon_B B \longrightarrow \upsilon_C C$ with the reaction rate corresponding to $r = k.[A]^n.[B]^m$

The E factor is estimated based on three dimensionless numbers defined below. The analysis of the E factor is presented in Table 5.3. The given key values used for these criteria are however subject to discussions and other values can be found in the literature (Doré, 1989, Charpentier and Wild, 1983).

The Hatta criterion: $$Ha = \sqrt{\frac{-\upsilon_B.k.[B].D_{A,l}}{k_l^2}}$$

The "concentration - diffusion" ratio: $$Z = \frac{\upsilon_A.D_{B,l}.[B]}{\upsilon_B.D_{A,l}.[A]^*}$$

The "reaction - transfer" ratio: $$R = \frac{k.[B].\varepsilon_l}{k_la}$$

We can now try to determine where the fastest reactions, assumed to be the "instantaneous" ozone demand, take place using some usual parameter values for static mixers:

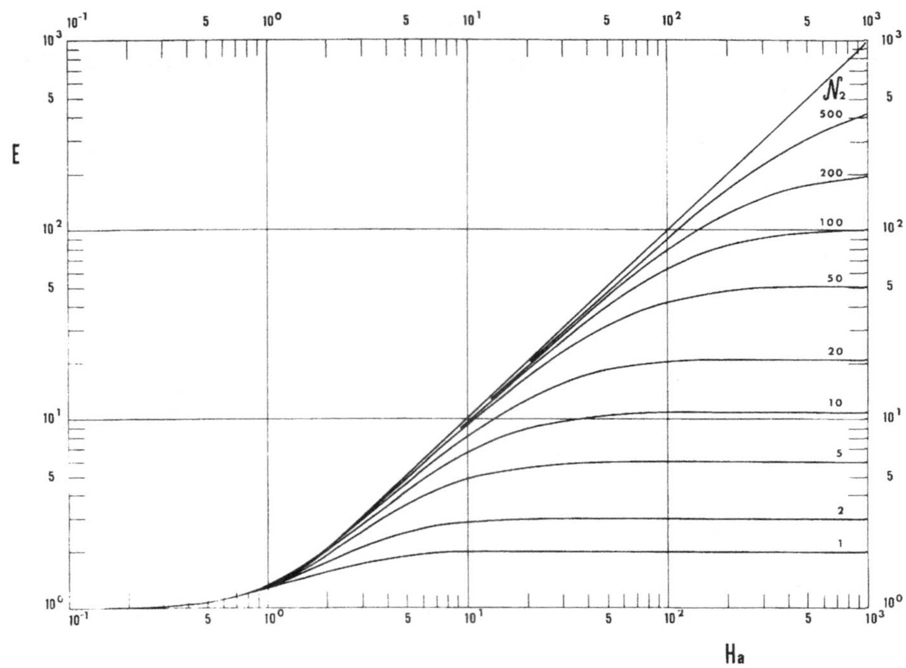

Figure 5.2 Abacus of Van Krevelen and Hoftijzer (1948)

$a \approx 1000\ m^2.m^{-3}$ $\quad K_La \approx 0.1\ s^{-1}$ $\quad K_L \approx 1\ 10^{-4}\ m.s^{-1}$

$D_{O3} = 1.71\ 10^{-9}\ m^2.s^{-1}$ at 15 °C $\quad \kappa \approx 1.5\ s^{-1}$

$$Ha = \sqrt{\frac{\kappa.D_{O_3,1}}{K_L^2}} = \sqrt{\frac{1.5 \cdot 1.71 \cdot 10^{-9}}{\left(1 \cdot 10^{-4}\right)^2}} = 0.506$$

We are therefore in a situation where the acceleration factor is slightly greater than one. The reactions take place simultaneously in the liquid film surrounding the bubbles but also in the bulk liquid. Hence, the linear profile approximation is no longer valid. And, if we need to estimate the mass transfer with the acceleration factor rate, more accurate information is needed to determine the ozone consumption in the film.

Static Mixer Modeling

From the two-film transfer model, the K_L value can be approximated by D_{O3}/δ where δ is the thickness of the diffusional layer. With the various correlations presented before, values for δ can be estimated between 5 and 50 μm according to velocities, holdup, type and geometry of the mixer.

If we assume that:

- The thickness is small compared to the diameter, i.e. the film can be considered as a flat sheet;

Table 5.3
The various situations of mass transfer with reaction (Martin and Laffort, 1991)

Chemical Reaction	Criteria	Concentration profiles	[A] value	E value	transfer equation	Main influence
Very slow	Ha < 0,3 R< 0,1	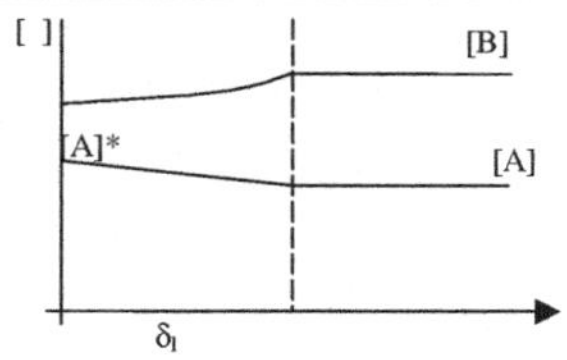	[A] # [A]*	E = 1	$dN=K_la([A]^*-[A])dV$ ou $dN = k[A]^*[B]dV$	Liquid volume
Slow	Ha < 0,3 0,1< R< 10	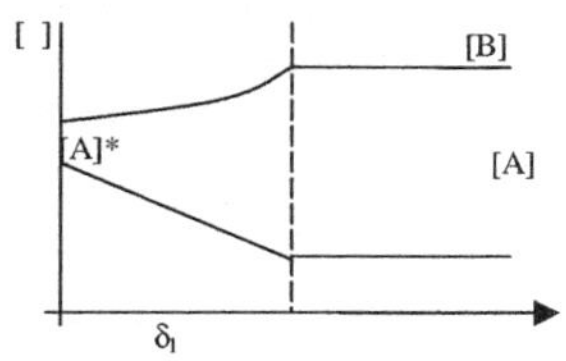	$0<N2<[A]^*$	E = 1	$dN=K_la([A]^*-[A])dV$	Liquid volume
Slow	Ha < 0,3 R> 10	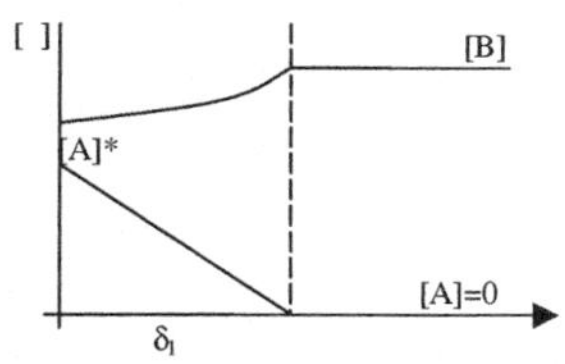	[A] # 0	E = 1	$dN=K_la[A]^*dV$	Interfacial area
Slightly fast	0,3< Ha <5	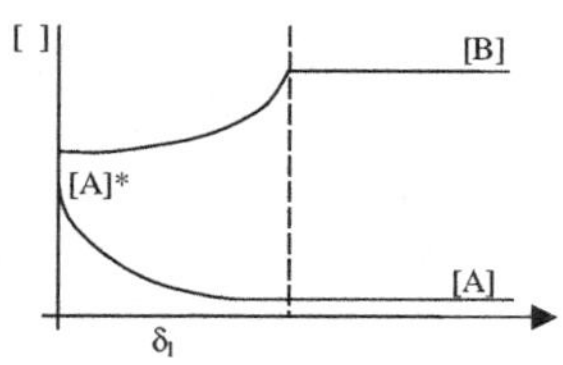	$0<[A]<[A]^*$	E=f(Ha,Z) Abacus of Van Krevelen	$dN=K_laE([A]^*-[A])dV$	Liquid volume and Interfacial area
Fast	Ha >5 Ha/Z <10	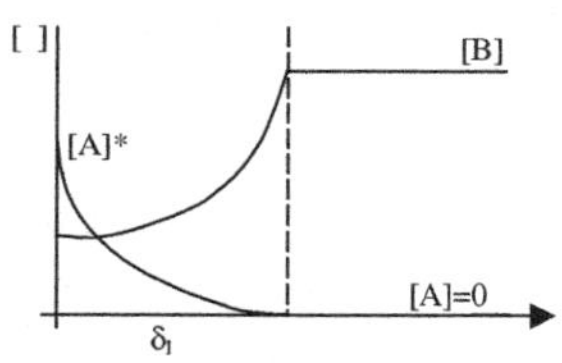	[A] = 0	E=f(Ha,Z) Abacus of Van Krevelen	$dN=K_laE[A]^*dV$	Interfacial area
Fast pseudo 1st order	Ha <5 $[B] >> [A]^*$	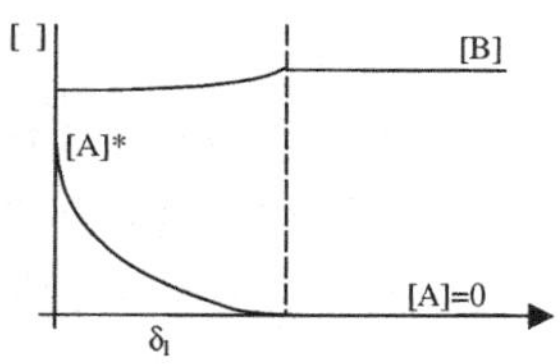	[A] = 0	E = Ha	$dN=k_l\,Ha\,[A]^*dV$ ou $dN=a[A]^*dV(k[B]D_l)_{0,5}$	Interfacial area
Instantaneous	Ha >5 Ha/Z >10	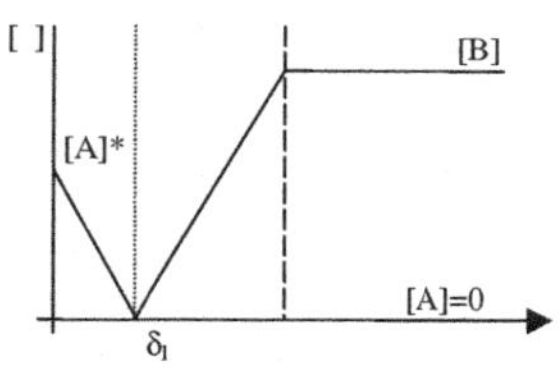	[A] = 0	E = 1+Z	$dN=k_laE[A]^*dV$	Interfacial area
At the surface	Ha >5 Ha/Z >10 $\frac{[B]}{C_g} < \frac{k_lD_l}{k_gD_B}$	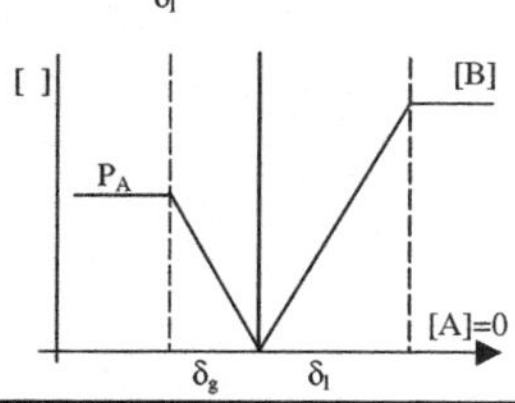	[A] = 0	-	$dN=k_la\,PdV$	Interfacial area Turbulence

- The bubble interface are stable, i. e. the same liquid fraction remains in direct contact with the gas all the way through the static mixer; and
- No transport and constant diffusion coefficient.

The mass balance of a film slice can be established:

$$\frac{\partial[X]_z}{\partial\tau} - D_X \frac{\partial^2[X]_z}{\partial z^2} - r_{X,z} = 0 \tag{5.20}$$

with X a compound concerned by the reaction mechanism.

τ, the bubble residence time coordinate

z, the thickness coordinate

r_X, the local production term

Such an equation requires boundary conditions:

1. At the gas - liquid interface (z = 0), the condition will depend on the type of compound:

   For the non volatile species, the diffusive potential must be zero:

$$\left.\frac{\partial[X]}{\partial z}\right|_{z=0} = 0 \tag{5.21}$$

   For volatile species, here we limit ourselves to ozone, the gas mass balance must be satisfied:

$$\frac{\partial n_{O_3,g}}{\partial\tau} = \frac{a}{\varepsilon - \varepsilon_g - a\cdot\delta}\cdot D_{O_3} \left.\frac{\partial[O_3]}{\partial z}\right|_{z=0} \tag{5.22}$$

   and the equilibrium must be satisfied:

$$[O_3]_{z=0,\tau} = [O_3]^*_\tau = \frac{\frac{P}{He}\cdot\frac{n_{O_3,g}}{\sum_i n_{i,g}}}{1-\frac{P}{He}\cdot\frac{n_{O_3,g}}{\sum_i n_{i,g}}}\cdot\left(\frac{\rho_{eau}\cdot 10^3}{M_{H_2O}}\right) \tag{5.23}$$

2. The second condition connects the boundary layer with the liquid bulk, i.e. the bubble and its surrounding with the rest of the liquid. The respective flow behavior will affect the mass transfer between the two. The flow pattern can be characterized based on the studies concerning the residence time distribution (Tung, 1976, Nauman, 1982, Nigam

and Vasudeva, 1980, Nauman and Vasudeva, 1985, Kemblowski and Pustelnik, 1988). These studies showed the flow induced by the static mixer can be considered as a perfect plug flow for all flow regimes (Chen et al., 1973, Middleton, 1985, Cybulski and Werner, 1986). We will therefore assume that liquid and bubbles progress "side by side" along the reactor. The liquid mass balance gives:

$$\left(\frac{\partial[X]_L}{\partial\tau} - r_{X,L}\right) = -\frac{a}{\varepsilon - \varepsilon_g - a\cdot\delta}\cdot D_X \cdot \left.\frac{\partial[X]}{\partial z}\right|_{z=\delta} \tag{5.24}$$

while a is the specific interfacial area relative to the total reactor volume, the ratio ($a/(\varepsilon-\varepsilon_g-a.\delta)$) is the specific interfacial area relative to the bulk volume.

3. Finally, initial conditions have to be defined. Following assumptions will be made:

At the entrance of the static mixer, the liquid film is a fraction of the bulk liquid

$$\text{for } z > 0 \quad [X]_z = [X]_L = [X]_{L,\tau=0} \tag{5.25}$$

Except for volatile compounds at the interface where the gas-liquid equilibrium is always satisfied :

$$[O_3]_{z=0,\tau=0} = [O_3]^*_{\tau=0} = \frac{\frac{P}{He}\cdot\frac{n_{O_3,g,\tau=0}}{\sum_i n_{i,g,\tau=0}}}{1-\frac{P}{He}\cdot\frac{n_{O_3,g,\tau=0}}{\sum_i n_{i,g,\tau=0}}}\cdot\left(\frac{\rho_{eau}\cdot 10^3}{M_{H_2O}}\right) \qquad n_{i,g,\tau=0} = \frac{Q_g\cdot[i]_{g,\tau=0}}{Q_L}$$

Up to now the bubbles were considered as being stable all through the static mixers. However, this seems doubtful since one of the major points in setting up static mixers (alternating elements and void spaces, partial rotation of the elements, etc.) is to create sudden obstacles which will divide the streamlines and modify their trajectory to generate turbulence and inertial impacts on the walls or the elements. It seems much more likely that the boundary will be periodically renewed, replaced by liquid from the bulk and some axial mixing does occur.

We can consider that the bubbles have an age probability distribution. We will assume an average normal distribution based on the structure of the mixer and define two parameters: the average age $\tilde{m}$ and the age variance σ^2.

$$P\left(m\in N\left(\tilde{m},\sigma^2\right)\right) = \frac{1}{\sqrt{\sigma^2\cdot 2\cdot\pi}}\cdot\exp\left(\frac{1}{2}\cdot\frac{(m-\tilde{m})^2}{\sigma^2}\right) \tag{5.26}$$

Once a bubble is destroyed and the film and the corresponding bulk mixed together, a new bubble would be created subjected to the same age distribution. In a perfect plug flow without axial mixing, this would lead to a very tedious calculation: if we divide the age distribution into 10 age groups and that 6 surface renewals occur during the crossing, this would require 10^6 different "stable boundary layer" simulations from 0 to $\tilde{m}$.

Now, if we accept some small local axial mixing, we could assume that the bubbles, and the corresponding liquid, of a same distribution are mixed together before generating new bubbles. So young bubbles and older bubbles would be mixed according to the expectancy given by the distribution. The jth mix formed will restart with a residence time equal to the cumulative average age of the original bubbles, and with age zero.

$$[X]_{\tau=j\cdot\tilde{m}} = \frac{1}{(\varepsilon - \varepsilon_g)} \int_0^{\infty} P\left(m \in N(\tilde{m}, \sigma^2)\right) \cdot \left(a \cdot \int_0^{\delta} [X]_{z,\tau=(j-1)\cdot\tilde{m}+m} dz + (\varepsilon - \varepsilon_g - a \cdot \delta) \cdot [X]_{L,\tau=(j-1)\cdot\tilde{m}+m} \right) \cdot dm$$

$$n_{O_3,g,\tau=j\cdot\tilde{m}} = \int_0^{\infty} P\left(m \in N(\tilde{m}, \sigma^2)\right) \cdot \left(n_{O_3,g,\tau=(j-1)\cdot\tilde{m}+m}\right) \cdot dm$$

As a result, a simulation with n rehomogenizations, is, more or less, equivalent to n+1 "stable boundary layer" simulations with an age overshoot to get the information for bubbles older than the average. For example, if a simulation time limit equal to $\tilde{m}$ + 1.96.σ is set: the oldest 2.5 % of the bubbles are omitted.

The reaction terms are obtained from the chemistry described earlier:

$$r_{O_3} = -\left(w + 2 \cdot k_2 \cdot 10^{pH-11.6} \cdot [H_2O_2] + \kappa + \sum k_{D,P_i} \cdot [P_i]\right) \cdot [O_3] \tag{5.27}$$

$$r_{H_2O_2} = -\left(k_2 \cdot 10^{pH-11.6} \cdot [O_3] + \left(k_6 \cdot 10^{pH-11.6} + k_7\right) \cdot [OH^\bullet]\right) \cdot [H_2O_2] \tag{5.28}$$

$$r_{P_i} = -\left(k_{D,P_i} \cdot [O_3] + k_{ID,P_i} \cdot [OH^\bullet]\right) \cdot [P_i] \tag{5.29}$$

The hydroxyl radicals are given by:

$$[HO^\bullet] = \frac{2 \cdot k_1 \cdot 10^{pH-14} + 2 \cdot k_2 \cdot 10^{pH-11.6} \cdot [H_2O_2] + K_D \cdot [TOC] + \beta \cdot \kappa}{(k_{11} + k_{12} \cdot 10^{pH-10.25}) \cdot [HCO_3^-] + K_{ID} \cdot [TOC] + \sum k_{ID,P_i} \cdot [P_i]} \cdot [O_3] \tag{5.30}$$

Gas/Liquid Separator

Several studies concerning the characterization and the use of ozone bubble reactors are available: Le Sauze (1990); Martin (1992); Wang (1995a,b); Ramirez (1997).

Mass Transfer Coefficients

For the mass transfer, we will limit ourselves to a brief summary of the overall mass transfer coefficient. Dudley (1995), Wang (1995a,b), Bin and Roustan (2000) summarized some of the knowledge about mass transfer parameters of gas/liquid separators. Numerous correlations have been established for the K_La parameter. For a specific system like the ozone/water system, the correlations can be simplified to $K_La = \alpha\ U_{s,g}^{\gamma}$ (Fuchum and Cunli, 1990). Several values can be found for various configurations (Roustan et al, 1996, Le Sauze, 1990, Fuchum and Cunli, 1990). For co-current flow, Wang (1995a,b) suggests:

$$\text{for } Re_L < 2000\ K_La\ (\text{min}^{-1}) = 0.092\ U_{s,g}^{0.516} \text{ with } U_{s,g} \text{ in m.h}^{-1}$$

and

$$\text{for } Re_L > 2000\ K_La\ (\text{min}^{-1}) = 0.110\ U_{s,g}^{0.615} \text{ with } U_{s,g} \text{ in m.h}^{-1}$$

Hydraulic Model

Because of the free circulation of both, the liquid and the bubbles, the flow pattern is relatively complex. The Navier and Stockes equations can be solved directly by computer simulations using codes like Fluent, Phoenics, etc. (Do-Quang et al. 2000). But the interactions of the two phases remain difficult to reproduce precisely and the introduction of the complex chemistry requires high performance computers.

Other models have been developed to describe the hydraulic behavior of such reactors (Villermaux, 1993, Roustan et al., 2000). The commonly used models are the axial dispersion plug flow model and the cascade of Continuous Stirred Tank Reactors model (Le Sauze, 1990; Martin, 1992, Marinas et al., 1993, Smith and Zhou, 1993, Roustan et al, 1994, Wang, 1995a,b). They were used for the description of the ozone profile in the reactor (Roustan et al., 1996, Le Sauze 1990, Orta de Velazquez, 1992, Wang, 1995a,b, Henry and Freeman, 1994, Reddy et al., 1995).

Several models can be considered, according to the flow behavior of each phase. A currently used model supposes a plug flow behavior of the gas phase and CSTRs for the liquid phase. The number of CSTRs can be determined with tracer tests. Table 5.4 gives some values obtained by Wang (1995a) for a co-current column (2.5 m height, 150 mm diameter).

Table 5.4
Number of CSTRs for a 2.5 m column (diameter : 0.15 m), co-current

		U_{sg} (m.h^{-1})		
	U_{sl} (m.h^{-1})	7.6	19.1	30.6
	16.95	492	491	478
τ (s)	47.46	177	171	171
	74.58	117	109	109
	16.95	2	2	2
J	47.46	5	4	3
	74.58	9	5	4

The mass balances for each stirred reactor can then be written:
For a non volatile compound in the liquid:

$$[X]_i - [X]_{i-1} = \frac{V_{L,i}}{Q_L} \cdot r_{X,i} \tag{5.31}$$

For a volatile compound (here ozone):

$$[O_3]_i - [O_3]_{i-1} = \frac{V_{L,i}}{Q_L} \cdot r_{O_3,i} + \frac{V_{L,i}}{Q_L} \cdot K_L a \cdot \Delta[O_3] \tag{5.32}$$

For the gas phase:

$$n_{O3,g,i} - n_{O3,g,i-1} = -\frac{V_{L,i}}{Q_L} \cdot K_L a \cdot \Delta[O_3] \tag{5.33}$$

with:

$$\Delta[O_3] = \left([O_3]_i^* - [O_3]_i\right) \text{ for a CSTR behavior for the gas phase} \tag{5.34}$$

or

$$\Delta[O_3] = \frac{\left([O_3]_i^* - [O_3]_i\right) - \left([O_3]_{i-1}^* - [O_3]_i\right)}{\ln\left(\frac{\left([O_3]_i^* - [O_3]_i\right)}{\left([O_3]_{i-1}^* - [O_3]_i\right)}\right)}$$

for a plug flow behavior of the gas phase (5.35)

In our case, the column has mainly a separation role. Most of the mass transfer has already occurred in the static mixers, the transfer in the column does essentially compensate for the ozone decay. There is little change in $[O_3]^*$ (ozone concentration at the interface), hence both values for $\Delta[O_3]$ are close.

The reaction terms are the same as the static mixer. Sometimes, if the amount of "instantaneous" ozone demand is not known, one can reasonably assume that this demand has been satisfied in the mixer and can be neglected in the gas/liquid separator, i.e. κ could be set to 0.

PROGRAM DESCRIPTION

The program simulates the transfer and the reactivity of ozone in natural waters in static mixers followed by a gas separation device. The information obtained at the mixer's outlet and at the column's outlet is:

- The average concentrations of each species: ozone, hydrogen peroxide, pesticide and hydroxyl radical in the liquid and in the gas for ozone;
- The ozone transfer efficiency;
- The fraction of transferred ozone consumed;
- The fraction of hydrogen peroxide consumed, when present; and
- The fraction of an organic compound (e.g. pesticides) degraded.

In addition for the static mixer, it gives the ozone, hydrogen peroxide (when present), pesticide and hydroxyl radical concentration profiles along the static mixer at several positions all through the boundary layer including the interface and the liquid bulk.

For the gas/liquid separator, it provides the concentrations in the intermediate CSTRs or plug flow sections.

Solution Methods

Static Mixer

The static mixer model is basically a set of two-dimensional partial differential equations (PDE) in time, or more exactly in bubble residence time, and in space (the layer thickness). The additional closure equations are satisfied by substitution in the main equations. For informative reasons, the algebraic equation relating the hydroxyl concentration to the ozone concentration is kept separate. The PDEs are of second order only in one direction. A discretization scheme according to the Crank-Nicholson implicit method can be used.

The time derivatives are expressed by forward first order approximation for each spatial discretization point:

$$\left.\frac{\partial [X]_z}{\partial \tau}\right|_{\tau}^{\tau+\Delta\tau} = \frac{[X]_z|_{\tau+\Delta\tau} - [X]_z|_{\tau}}{\Delta\tau} \tag{5.36}$$

The second order space derivatives are obtained by central first order approximation using the time average concentration values at each spatial discretization point.

$$\left.\frac{\partial^2 [X]_z}{\partial z^2}\right|_{z}^{z+\Delta z} = \frac{[\tilde{X}]_{z+\Delta z} - 2\cdot[\tilde{X}]_z + [\tilde{X}]_{z-\Delta z}}{\Delta z^2} \tag{5.37}$$

with $[\tilde{X}]_z = \dfrac{[X]_z|_{\tau} + [X]_z|_{\tau+\Delta\tau}}{2}$

The reaction terms are determined for the time average concentrations.

Some adjustments are needed for the discretization points corresponding to the boundary conditions, i.e. the gas-liquid interface and the liquid bulk.

According to these choices, we obtained a set of non-linear algebraic equations to be solved for each new time step: we know the [X] profile at τ and want to determine the [X] profile at $\tau+\Delta\tau$. These equations need to be solved simultaneously since the solution at z requires values for $z-\Delta z$ and $z+\Delta z$. The relatively simple expression of the various equations makes it possible to use efficiently a Newton-Raphson method taking advantage of the sparse Jacobian matrix. Stating from initial values, this method simply assumes a first order linear approximation in order to determine a correction for all the variables. The new values are used to check the satisfaction of the residual functions and, if no satisfaction is obtained, they're used to generate a new first order linear approximation.

$$F([X]+\Delta[X]) \approx F([X])+\nabla^T_{[X]}F \cdot \Delta[X]=0$$

$$\Delta[X]=-\left(\nabla^T_{[X]}F\right)^{-1} \cdot F([X])$$

$$[X]=[X]+\Delta[X]$$

$$F([X])=0 \quad ?$$

The size of the time step and the space step have been determined by successive simulations with, each time, smaller values until their was no more of a significant change in the resulting information: at least 200 space intervals and $\Delta\tau$ in the 10^{-4} s range.

For the surface renewal equations and the average liquid concentration, the trapezoidal rule is used for the integration over space and over bubble ages.

Figure 5.3 gives a coarse description of the program structure for the static mixer model.

Gas/Liquid Separator

The CSTR cascade model corresponds to a small set of algebraic non-linear equations which can be solved easily by a Newton-Raphson method.

Figure 5.4 gives a coarse description of the program structure for the CSTR model.

Software Structure

The program has been written using the FORTRAN 77 programming language. The 27 subroutines form a total of 2600 lines. The user interacts with the program through data and result files. Three files gather the needed information:

- A file containing the operating conditions of the reactor;
- A file containing the kinetic data; and
- A file containing some parameters for the numerical methods.

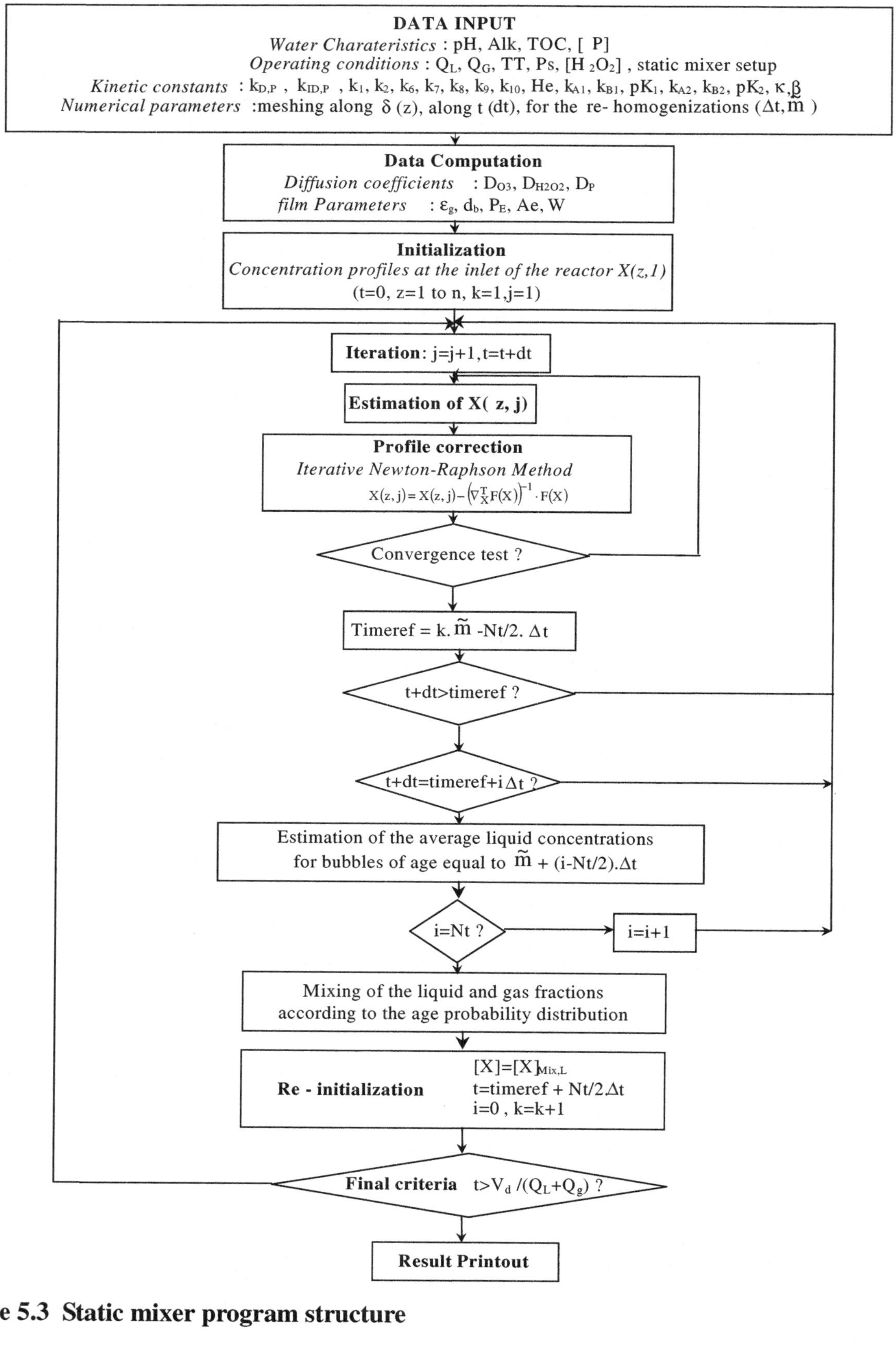

Figure 5.3 Static mixer program structure

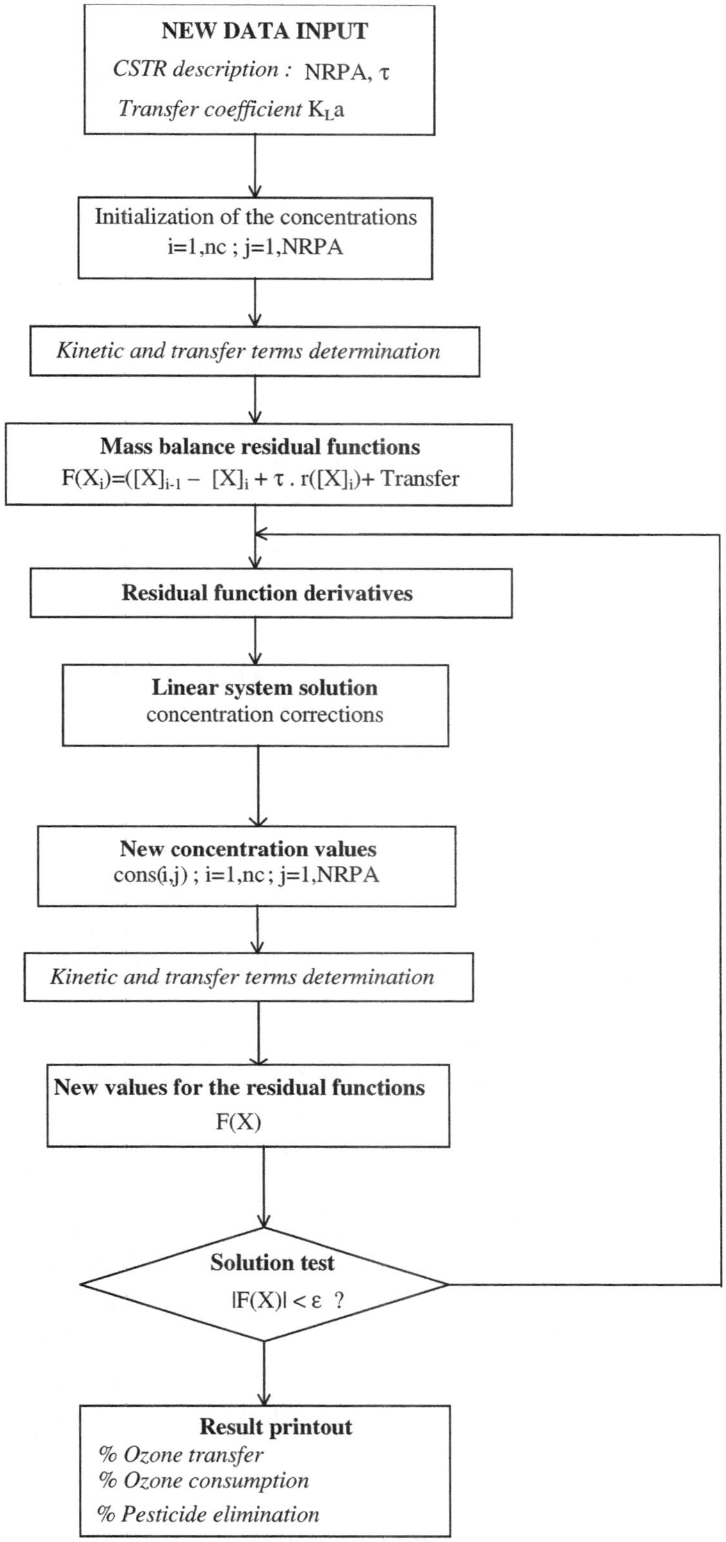

Figure 5.4 Gas/liquid separator sub-program structure

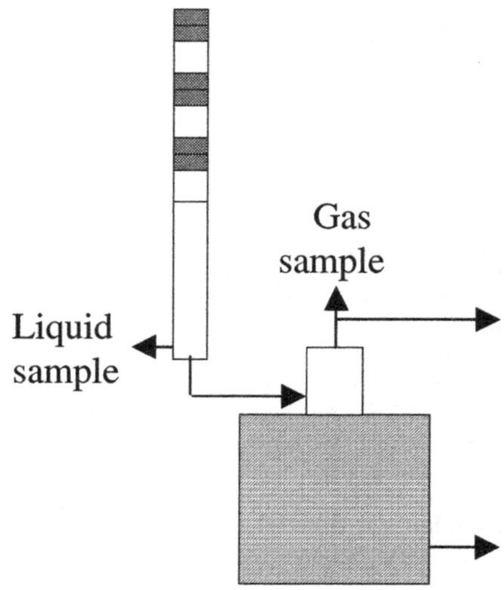

Figure 5.5 The lab-scale unit at the University of Rennes

EXPERIMENTS, SIMULATION AND MODEL VALIDATION

Laboratory Tests

The Lab-Scale Unit at the University of Rennes

This lab unit was set up and operated by Langlais (1998) (see Figure 5.5). A vertical downstream static mixer with six elements (Sülzer SMV4 - 20 in the optimal configuration) in terms of transfer and energy consumption was established by Hérès (1993).

The modules are assembled two by two with a relative 90° rotation with an equivalent void space between each pair. An equally long empty pipe follows the mixers before a liquid sample can be taken. The mixture enters then, slightly above the water level, a large tank with a relative small headspace for the separation of the gas and the liquid phase and the collection of a gas sample.

The characteristics of the lab-scale unit are given in Table 5.5.

The experiments were conducted with tap water from the City of Rennes. The water was spiked with atrazine for which decay can be estimated based simply on the ozone and the hydroxyl radical concentrations. It gives therefore a tool for the validation of the hydroxyl radical concentration. The varying parameters were: gas flowrate, liquid flowrate, and ozone dose. The collected laboratory data were the ozone concentrations at the inlet and the outlet of the static mixer unit in both gas and liquid phases, as well as the atrazine concentrations in the liquid. The efficiency of the static mixer was determined based on:

- The ozone transfer ratio:

$$\%T = \frac{n_{O_3,0} - n_{O_3,\tau}}{n_{O_3,0}} = \frac{Q_{g,0} \cdot [O_3]_{g,0} - Q_{g,\tau} \cdot [O_3]_{g,\tau}}{Q_{g,0} \cdot [O_3]_{g,0}}$$

Table 5.5
Rennes' lab scale unit geometrical data

Part	Parameter	Value (SI units)	Value (US units)
Stainless steel pipe	Internal diameter	22.3 mm	0.87 inch
	Length	130 mm	5.08 inches
Mixing element (3/4")	Length	20 mm	0.78 inch
SMV 4	Hydraulic diameter	5.0 mm	0.197 inch
	Void fraction	0.87	0.87
Reactor volume		$8.76\ 10^{-5}\ m^3$	$2.31\ 10^{-2}$ gal

- The consumed fraction of transferred ozone:

$$\%Cons = \frac{n_{O_3,0} \cdot \%T - [O_3]_\tau}{n_{O_3,0} \cdot \%T}$$

- The atrazine oxidation efficiency:

$$\%Atz = \frac{[Atz]_0 - [Atz]_\tau}{[Atz]_0}$$

Following parameters from the literature have been used:

- $\kappa = 1\ s^{-1}$, this order of magnitude of "instantaneous" ozone demand of the water for the City of Rennes has been evaluated by Munoz Ramirez (1997).
- $k_{OH°,P}=6.6x10^9\ L.mol^{-1}.s^{-1}$, which is an average value from the literature.
- $K_D=0.2$ et $K_{ID}=1.58x10^6$, have been estimated for this water using the values given by Perrot (2000) for the Neuilly sur Marne water.
- The ozone kinetics by Wang will be used because this model is based on tap water data.
- $K_La=0.03\ s^{-1}$, the volumetric global liquid mass transfer coefficient for the CSTRs. This value is usual according to the literature.
- 2 CSTRs are sufficient to model the empty pipe length because of the small volume.
- 8 film age groups were sufficient for each re-homogenization (Bais, 2000).

Some parameters needed to be adjusted:

- β = 0.8 is the proportionality coefficient between the ozone consumed by the instantaneous demand and the hydroxyl radicals produced by these reactions. This value has been required to obtain a satisfying adjustment of the atrazine degradation efficiency, confirming that the organic matter responsible for the instantaneous demand induces also a radical formation.
- 6 re-homogenizations were used within the static mixer.

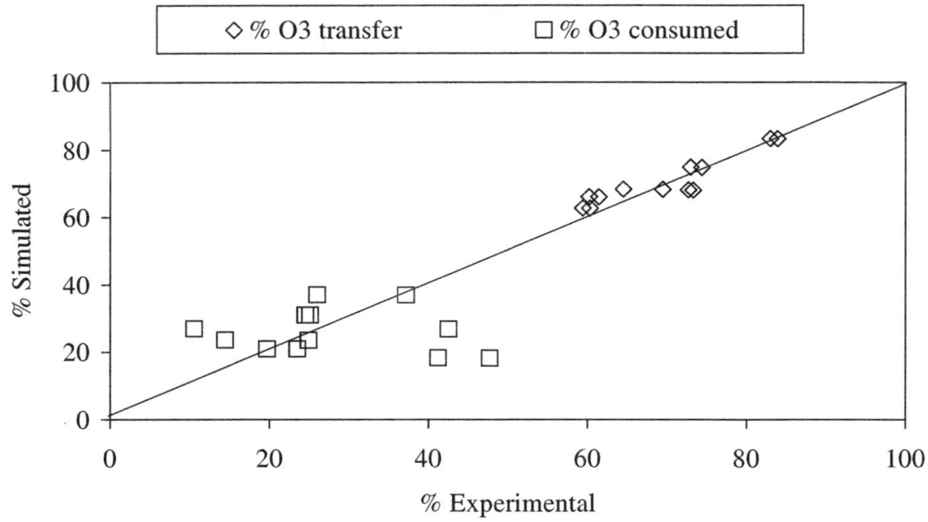

Figure 5.6 Simulation of ozone transfer and consumption – Lab-scale testing

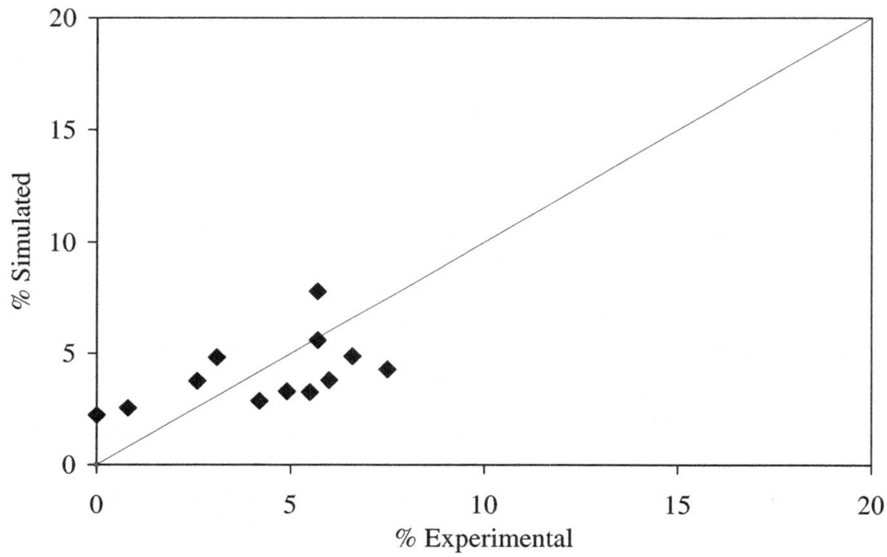

Figure 5.7 Simulation of atrazine removal – Lab-scale testing

Figures 5.6 and 5.7 compare the simulated results to the laboratory data. The mass transfer efficiency can be estimated with an absolute error of about ± 5 %. The atrazine removal has an absolute error of ± 3 %, which is somewhat equal to the laboratory error for such small removal rates.

The ozone consumption ratio is less satisfying: while a close average value for the consumption could be obtained, there is a significant dispersion. A reason for this could be that the various manipulations were not done the same day. Usually, since the main objective was the advanced oxidation process, the ozone alone experiment was just the first of a set of experiments with increasingly more hydrogen peroxide. The water characteristics may have been slightly different for each day, but these differences, having no major consequences for advanced oxidation, may not have been closely monitored.

The developed simulation program, based on the model described earlier, reproduces correctly the laboratory data obtained by Langlais (1998) on tap water. However, the reliability of the model must be confirmed by challenging it to various other situations: other waters, other static mixer setups, other sizes, etc.

Pilot-Scale Units

The French Pilot Unit

The French pilot unit was described in Chapter 3 under the Section "French Pilot Study". The dissolved ozone concentration was measured at the outlet of the static mixer and of the gas/liquid separator. The ozone gas concentration was only measured at the outlet of the column.

Following parameters from the literature were used:

- The exact "instantaneous" ozone demand of the water for the Neuilly-sur-Marne water plant is not known: Munoz Ramirez (1997) found values between 0.3 and 1.4 $mgO_3.L^{-1}$ over a six month period for this water and an average $\kappa = 1\ s^{-1}$. In the program, the kinetic rate constant is given and "instantaneous" ozone reaction is expected to be complete at the outlet of the mixer. As a result, if we want to maintain the amount or instantaneous demand needs to be maintained when the mixer volume increases, the kinetic constant has to be reduced. Thus $\kappa = 1.5\ s^{-1}$ for 2 SM; $\kappa = 1\ s^{-1}$ for 4 SM; and $\kappa = 0.6\ s^{-1}$ for 7 SM.
- $k_{OH°,P}=6.6x10^9\ L.mol^{-1}.s^{-1}$, which is an average value from the literature.
- $K_D=0.2$ et $K_{ID}=1.58x10^6$, were obtained for this water by Perrot (2000).
- The ozone kinetics by Orta de Velasquez (1992) will be used because this model concerns settled waters.
- $K_La=0.03\ s^{-1}$, the volumetric liquid mass transfer coefficient for the CSTRs. This value is usual according to the literature.
- The column is represented by 4 CSTRs. The residence time is derived from the effective height and the liquid and gas flowrate.
- Since no radical indicator has been introduced, the proportionality coefficient between the ozone consumed by the instantaneous demand and the hydroxyl radicals produced by these reactions has been estimated as same as before: $\beta = 0.8$

Some parameters needed to be adjusted :

- 6 to 8 re-homogenizations were used within the static mixer depending on the number of static mixer modules.

Figure 5.8 shows the comparison of the laboratory and simulated ozone transfer and consumption for the static mixer and degassing column, while Figure 5.9 and Figure 5.10 represent the comparison of the laboratory and simulated ozone concentration at the outlet of the static mixer and the degassing column.

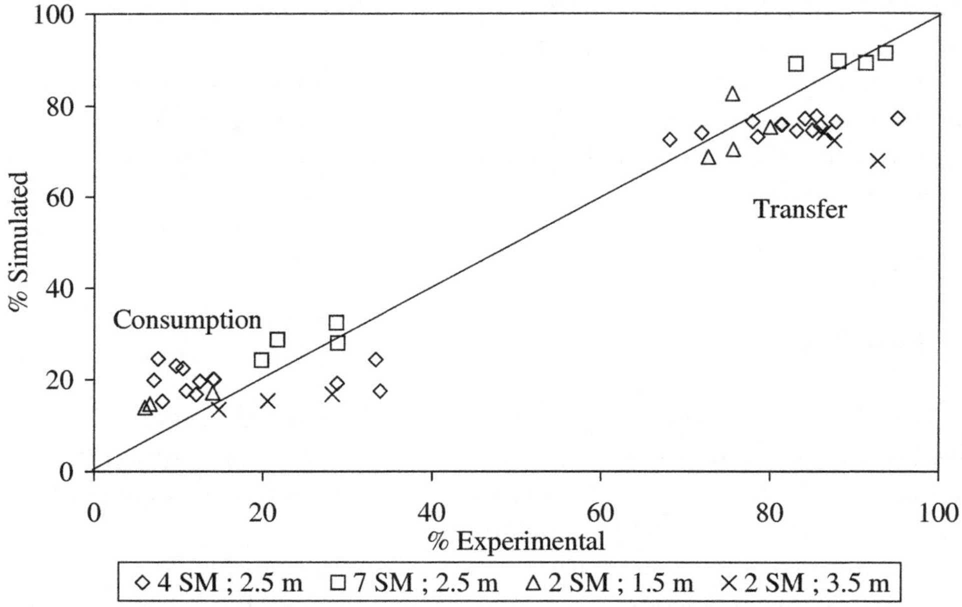

Figure 5.8 Simulation of ozone transfer and consumption – French pilot testing

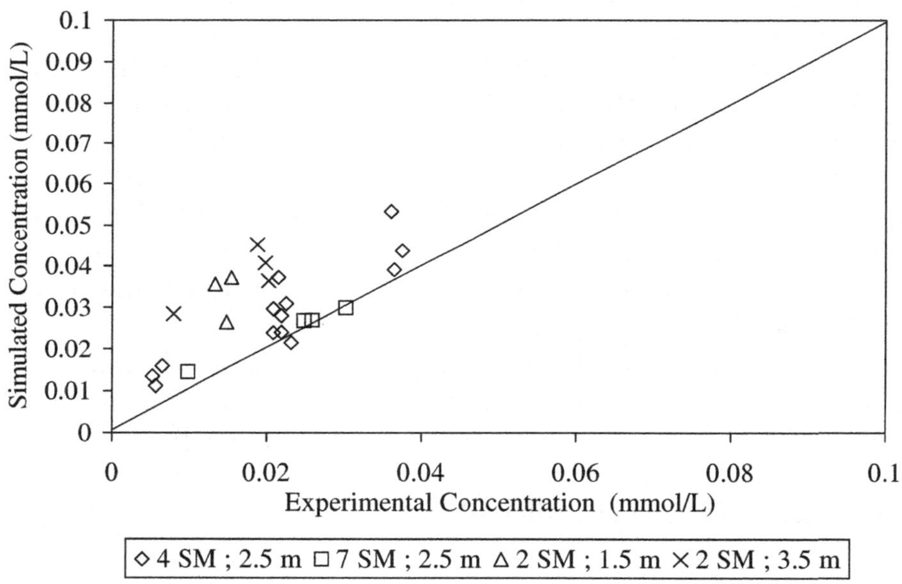

Figure 5.9 Simulation of ozone residual at the outlet of the static mixer – French pilot testing

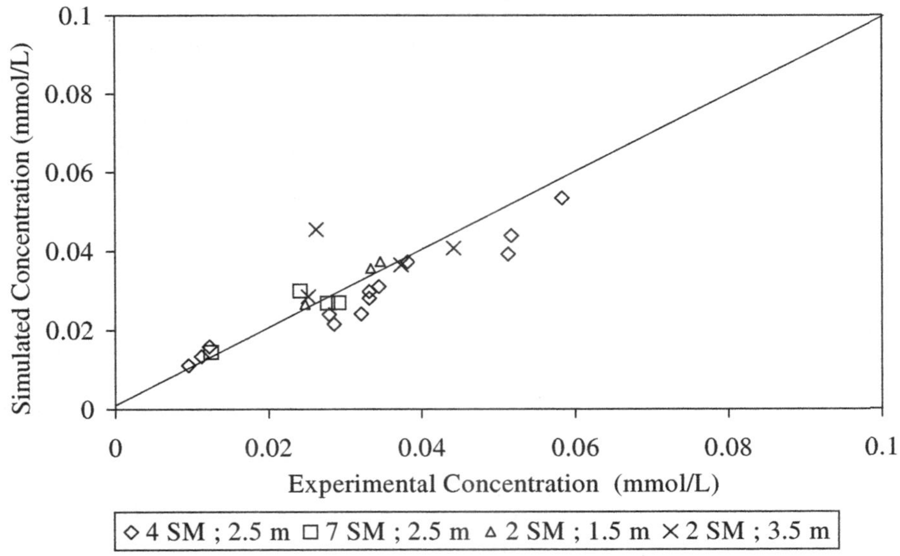

Figure 5.10 Simulation of the ozone residual at the outlet of the degassing column – French pilot testing

These results are quite satisfactory, especially for the ozone transfer and the final ozone concentration. For the ozone concentration at the outlet of the static mixer, the results appear less correlated mainly because of the values for the 2 module static mixer experiments.

The model does correctly simulate the transfer and reactivity of ozone in static mixers for various configurations, as well as the contribution of the gas/liquid separator.

Pilot Studies in the U.S.

The description of the pilot unit can be found in Chapter 3 under the Section "United States and United Kingdom Pilot Studies". Pilot-scale results from the following demonstration sites were used:

- raw and filtered water from the Gwinnett County water plant, Georgia (GC);
- raw water from Cleveland, Ohio (CWD);
- raw water from the City of Elizabethtown, New Jersey (EWC);
- raw water from the Los Angeles Department of Water and Power, California (LADWP); and
- filtered water prior to the intermediate ozonation step at the Chertsey WTP (CWTP).

The following parameters were used:

- $K_La = 0.03\ s^{-1}$, the volumetric overall liquid mass transfer coefficient for the CSTRs. This value is widely used in the literature.
- The ozone decay kinetics developed by Orta de Velasquez (1992) was used because this model was the closest to those types of water.

Table 5.6
Kinetic parameters for the US/UK pilot experiments

	Unit	Chertsey	Cleveland	Elizabethtown	Gwinnett	Los Angeles
kA_1	$L.mg^{-1}.s^{-1}$	0.0	0.0	0.0	0.0	0.0
kB_1	$L.mg^{-1}.s^{-1}$	0.002	0.04	0.005	0.04	0.04
pK_1	-	0.0	0.0	0.0	0.0	0.0
kA_2	$L.mg^{-1}.s^{-1}$	0.0	0.0	0.0	0.0	0.0
kB_2	$L.mg^{-1}.s^{-1}$	1.58E6	1.58E6	1.58E6	1.58E6	1.58E6
pK_2	-	0.0	0.0	0.0	0.0	0.0
κ	s^{-1}	3.0	1.0	7.0	2.5	2.5
β	-	0.8	0.8	0.8	0.8	0.8
Model (icod)	-	0	2	0 - 2	2	0
K_La	s^{-1}	0.03	0.03	0.03	0.03	0.03

- The column was overall represented by 4 CSTRs. The residence time was derived from the effective height and the liquid and gas flowrate. The plug flow section can be represented by a large series of small CSTRs.
- K_D=0.2 and K_{ID}=1.58x10^6, obtained by Perrot (2000).
- The proportionality coefficient between the ozone consumed by the instantaneous demand and the hydroxyl radicals produced by these reactions was assumed similar for each water: $\beta = 0.8$
- In order to fit the ozone concentration at the outlet of the static mixer, the amount of "instantaneous" ozone demand of the water was adjusted to $\kappa = 0.3\ s^{-1}$.

Different kinetic parameters were sometimes used for each water plant to take into account the variability of the water treated (Table 5.6). However, all the experiments at a same location use the same parameters. Mainly affected were the instantaneous ozone demand and the production of hydroxyl radicals from the reaction between ozone and the organic matter.

The value of kA and pK for the reactions between ozone or hydroxyl radicals and the organic mater for the hydroxyl radical molar balance have been chosen so as to reduce to a single constant:

$$K_D = \frac{k_{A1} \cdot 10^{pK_1 - pH} + k_{B1}}{1 + 10^{pK_1 - pH}} = \frac{0 \cdot 10^{0-pH} + k_{B1}}{1 + 10^{0-pH}} \cong k_{B1} \tag{5.38}$$

Figures 5.11, 5.12, 5.13 and 5.14 show the agreement between experimental and simulated transfer efficiency, ozone consumption rate and ozone residual after the gas/liquid separator.

Overall, a good agreement was obtained with the available data. Of course, the use of adjusted parameters should be reduced to the minimum for the design of transfer units.

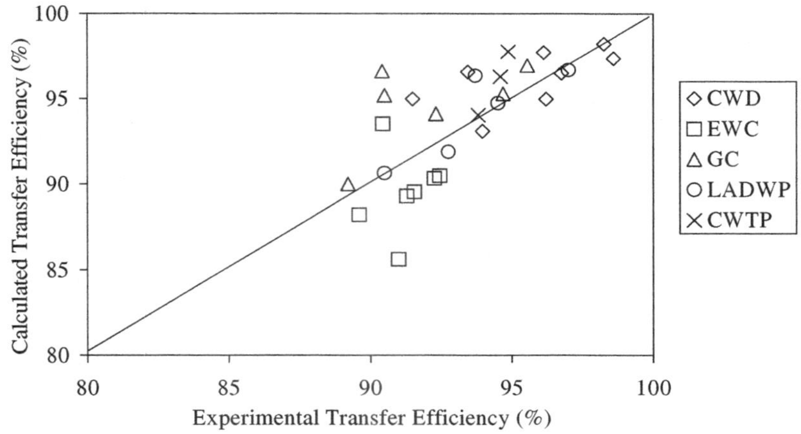

Figure 5.11 Simulated versus actual ozone transfer efficiency – US pilot testing

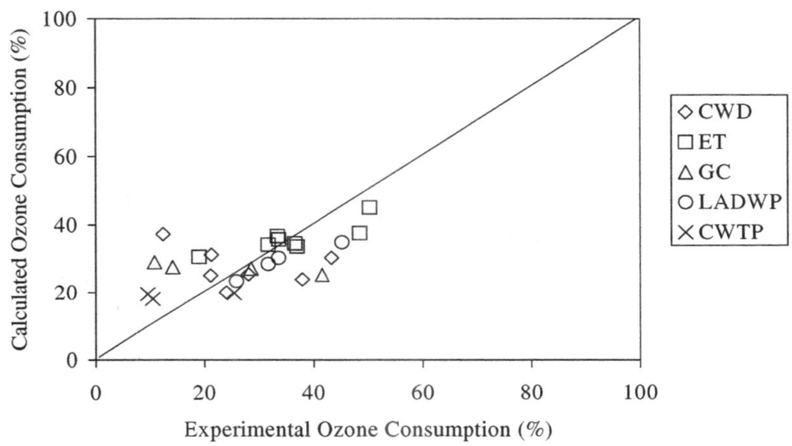

Figure 5.12 Simulated versus actual ozone consumption ratio – US pilot testing

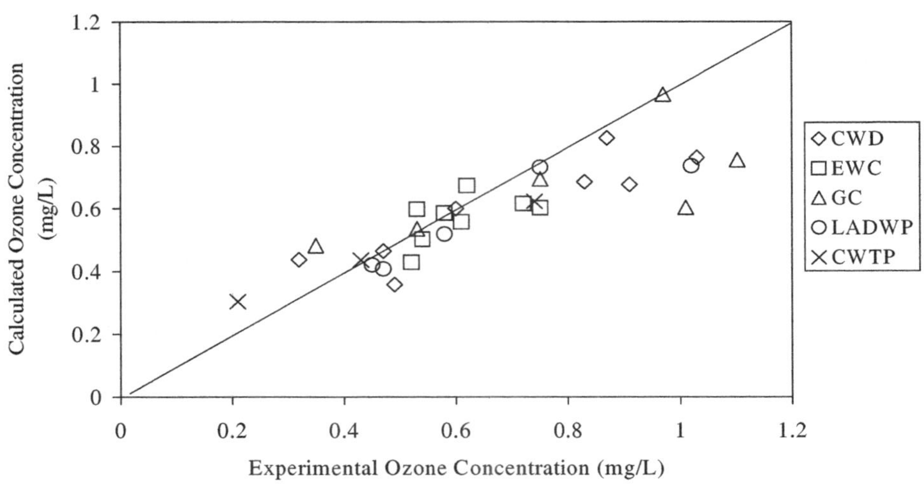

Figure 5.13 Simulated versus actual ozone residual at the outlet of the static mixer – US pilot testing

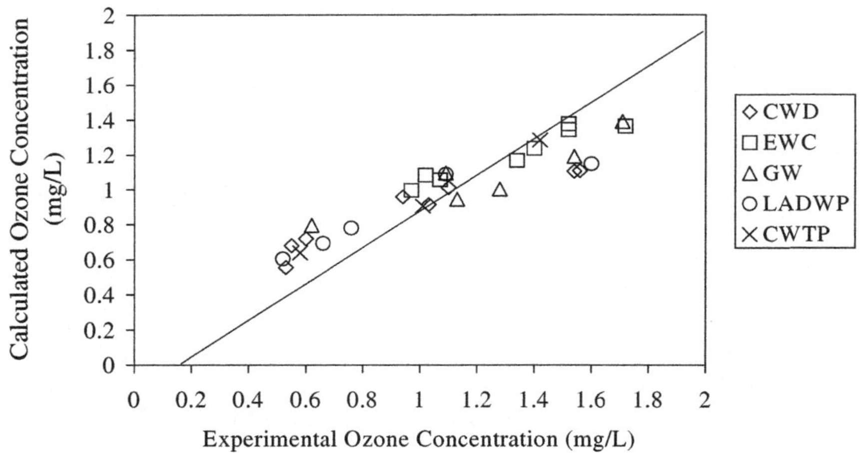

Figure 5.14 Simulated versus actual ozone residual at the outlet of the gas/liquid separator – US pilot testing

SUMMARY

The model, and the corresponding program, has been tested over a large set of laboratory and pilot data: lab-scale experiments carried out at University of Rennes, pilot-scale experiments at the Neuilly sur Marne water treatment plant, pilot-scale experiments in the U.S. and in the U.K. For all these examples, a satisfactory simulation could be obtained for the ozone concentration (residual) at the outlet of the static mixers or of the gas/liquid separator device, ozone consumption rate and for the transfer efficiency. The model did correctly simulate the transfer and reactivity of ozone in static mixers for various configurations, as well as the contribution of the gas/liquid separator.

CHAPTER 6
BROMATE MODELING

INTRODUCTION

Apart from the potential attributes of using a static mixer in terms of enhanced ozone mass transfer and dissolution, a major question facing any ozone application is potential and amount of bromate (BrO_3^-) formation. The goal is to reconcile effective ozone disinfection (reduction of microbial risk) while minimizing bromate formation (chemical risk). While the goal of disinfection is elimination of a target microorganism, disinfection practice is based on Ct (mg/L-min) requirements to achieve a potential log inactivation corresponding to some combination of ozone concentration/residual, C (mg/L), and contact/residence time, t (min).

The presence of natural organic matter (NOM), measured as dissolved organic carbon (DOC), exerts an ozone demand and thus necessitates a higher ozone dose to provide a given Ct. The provision of a Ct requirement is also influenced by the specific type of ozone dissolution system employed. A typical fine bubble diffuser (FBD) contactor consists of one or more dissolution cells followed by one or more residence/reaction cells, with ozone application provided by diffusers situated at the bottom of application cells. Calculation of the Ct for the overall contactor is based on the summation of individual-cell Ct estimates, excluding the first (application) cell. The precise interpretation of C corresponds to the cell effluent, C_{effl}, while t is precisely interpreted as the 10th percentile value, t_{10}. In a static mixer (SM) system, the diffuser is replaced by the SM followed by one or more residence/reaction cells.

BROMATE CHEMISTRY

In the presence of a dissolved ozone residual, bromide ion (Br^-) is oxidized to bromate according to either a molecular ozone (O_3) and/or hydroxyl radical ($OH\cdot$) pathway(s):

Molecular Ozone Pathway:
$Br^- + O_3 \rightarrow OBr^- + O_2$ ($HOBr \leftrightarrow H^+ + OBr^-$)
$OBr^- + O_3 \rightarrow BrO_3^- + ½ O_2$

Hydroxyl Radical ($OH\cdot$) Pathway:
$Br^- + OH\cdot \rightarrow BrO_3^-$

Influential factors and their relative influence (+ or -) on bromate formation include: O_3 dose (+), Br^- (+), pH (+), ammonia-nitrogen (-), carbonate alkalinity (+), temperature (+), DOC (+/-) and contact/residence time (+). The formation of bromate can be (chemically) minimized by acid (pH depression) or ammonia addition.

Bromate Predictive Models

Part of the overall SM database (four sources with robust databases; Cleveland, UK, Elizabethtown, and LADWP) was used to develop a bromate formation model for the static mixer pilot unit, designated as the SMU model and shown below; parameter units are BrO_3^-, μg/L; O_3, mg/L; Br^-, µg/L; temperature (temp), °C; DOC, mg/L; and alkalinity(alk), mg/L. This model does not incorporate the contact time as variable as it was kept constant for each test (contact time ≈ 25 minutes). Therefore, the model can only be used to predict bromate formation for a contact time of 25 minutes.

$$BrO_3^- = 5.81\times10^{-4}(O_3)^{1.31}(Br^-)^{0.49}(temp)^{-2.11}(pH)^{2.45}(DOC)^{-1.88}(alk)^{2.00} \quad (6.1)$$

The FBD unit database was not robust enough (only two sources) to consider developing a corresponding predictive model. However, a previously developed model was used to simulate bromate formation in a FBD ozone contactor, designated as the Integrated Disinfection Design Framework (IDDF) model (Carlson et al., 2001) and shown below. Parameter units are the same as the SM model and note that this model does not include *alkalinity* as a variable. Table 6.1 summarizes the definitions and boundary conditions of the IDDF model.

$$BrO_3^- \text{ @ } 20\ °C = 1.55x10^{-6}(O_3)^{1.57}(Br^-)^{0.73}(pH)^{5.82}(DOC)^{-1.26}(t)^{0.28} \quad (6.2)$$

$$BrO_3^- \text{ @ temp} = BrO_3^- \text{ @ } 20\ °C(1.035)^{temp-20} \quad (6.3)$$

A series of model simulations are shown in Figures 6.1 and 6.2, representing scatter plots of predicted versus measured bromate values, based on three modeling scenarios: (i) the SMU model; (ii) the FBD (IDDF) model simulating contactor hydraulics of t_{10}/t_{50} = 1.0, corresponding to a plug flow reactor; and (iii) the FBD (IDDF) model simulating more typical FBD hydraulic conditions of t_{10}/t_{50} = 0.7. Since only data derived from the SM pilot unit were used in SMU model calibration, the SMU model embodies its inherent hydraulic (i.e., t_{10}/t_{50}) conditions. The SMU model (R^2=0.95) provided better predictive capabilities than the FBDU (IDDF) model (R^2=0.62-0.70). The FBD (IDDF) model/modeling scenarios under-predicted bromate formation and this is probably due to the fact that it does not take into account the alkalinity of the water into consideration.

The SMU and FBD (IDDF) models were then used to make sets of parallel predictions based on common input data sets (Cleveland, UK, Elizabethtown and LADWP) corresponding to the respective testing conditions (O_3, Br^-, pH, DOC, t, temperature, and alkalinity (SMU only)). These simulations are presented in Figures 6.3 and 6.4.

Table 6.1
Boundary conditions for the IDDF bromate model

Parameters	Definition (units)	Boundary conditions
DOC	DOC (mg/L)	1.1 to 8.4
Br^-	Bromide (µg/L)	70 to 440
pH	pH	6.5 to 8.5
temp	Temperature (°C)	20
O_3	Transferred/utilized ozone (mg/L)	1.1 to 10.0
t	Time (min)	1 to 120
BrO_3^-	Bromate (µg/L)	2 to 314

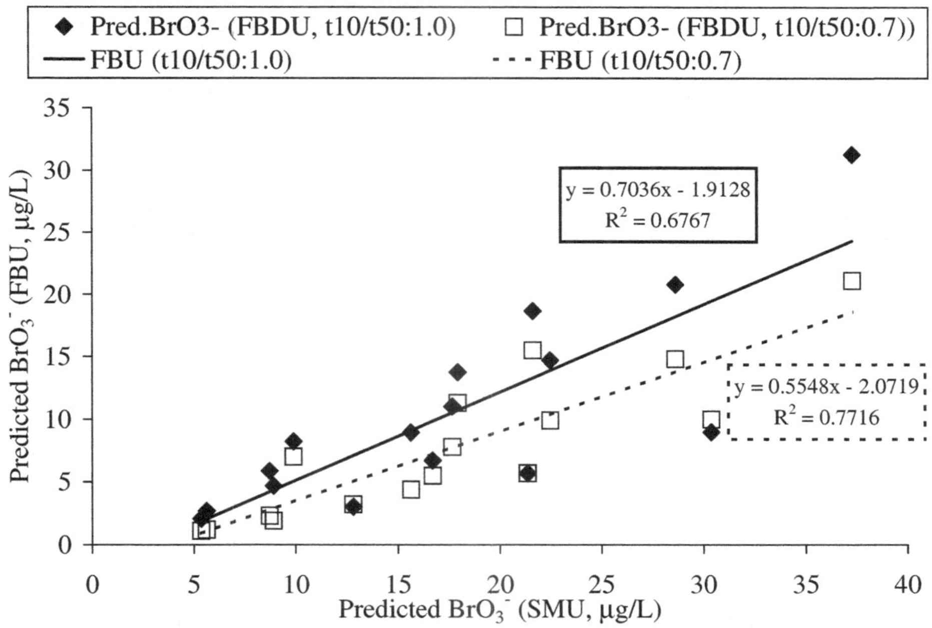

Figure 6.1 Comparison of SMU model and IDDF model for bromate formation at actual temperature

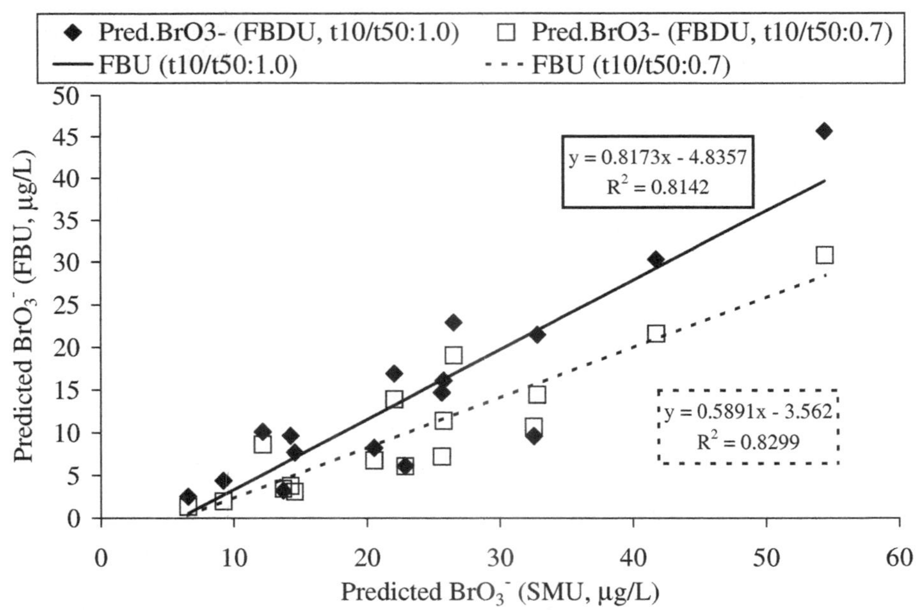

Figure 6.2 Comparison of SMU model and IDDF model for bromate formation at 20°C

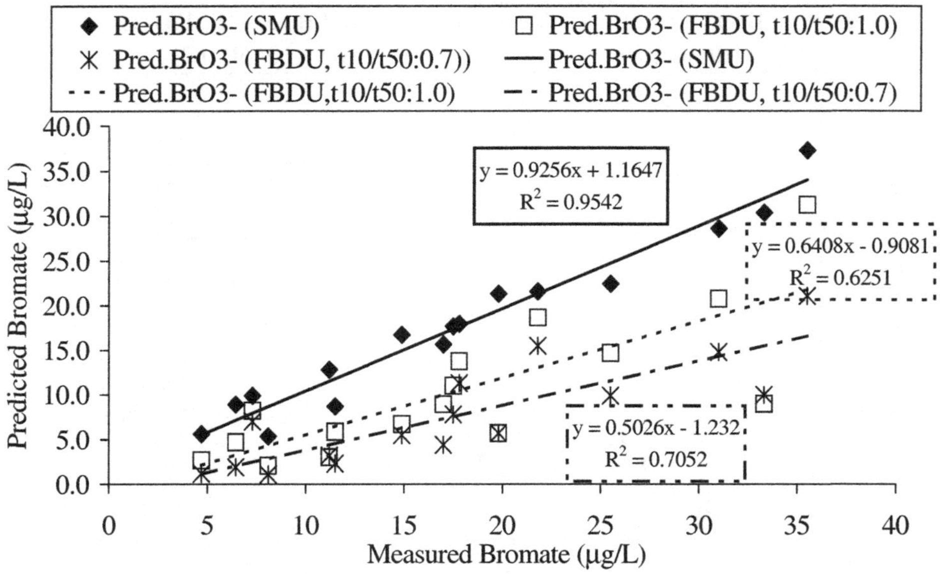

Figure 6.3 Evaluation of SMU model (Cleveland, UK, Elizabethtown, and LADWP) and IDDF model (FBU, t_{10}/t_{50}=1.0 and 0.7) for bromate formation at actual temperature

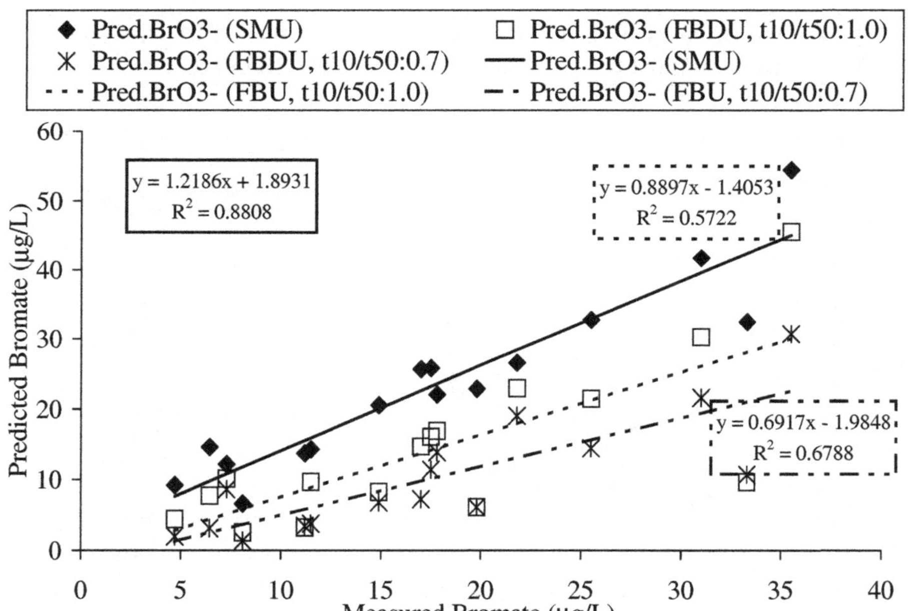

Figure 6.4 Evaluation of SMU model (Cleveland, UK, Elizabethtown, and LADWP) and IDDF model (FBU, t_{10}/t_{50}=1.0 and 0.7) for bromate formation at 20°C

These simulations suggest that significantly lower bromate formation is predicted for the FBD unit than measured values. The FBD (IDDF) model (simulating more typical FBD hydraulic conditions of $t_{10}/t_{50} = 0.7$) predicts lower bromate formation than the ideal FBDU (IDDF) model (simulating contactor hydraulics of $t_{10}/t_{50} = 1.0$). Similar (at actual temperature) or slightly higher bromate (at 20°C) compared to measured values were predicted for the SM unit, under comparable water quality and ozonation (dose and contact time) conditions.

CONCLUSIONS

Bromate formation models were developed incorporating the following parameters: bromide concentration, ozone dose, pH, temperature, DOC, contact time and alkalinity. Based on bromate versus global Ct trends for four sources (Cleveland, UK, Elizabethtown, and LADWP), there appears to be no clear benefit in the use of the SM over the FBD system, or vice versa, in terms of bromate minimization. Side-by-side studies at Elizabethtown and LADWP indicated that there is no consistently-uniform, compelling evidence to suggest that the use of a SM system versus FBD system will have a *significant* effect, enhancement or diminishment, on bromate formation; and thus, other criteria (economics, footprint, better microorganism inactivation, ease of operation and maintenance) will weigh more heavily in process/system selection.

The simulations suggested that significantly lower BrO_3^- formation is predicted for the FBD unit than measured values. This may be due to the fact that these models do not incorporate the influence of alkalinity. The SMU model provided better predictive capabilities than the FBD (IDDF) model. Similar (at actual temperature) or slightly higher BrO_3^- (at 20°C) compared to measured values were predicted for the SM unit, under comparable water quality and ozonation (dose and contact time) conditions.

CHAPTER 7
CONCLUSIONS AND RECOMMENDATIONS

OVERVIEW

This report presents the results of an in-depth, comprehensive study on the use of static mixers for ozonation in drinking water treatment. The results obtained from lab-scale (Chapter 2); pilot-scale (Chapter 3); and full-scale studies (Chapter 4) over a wide variety of operating conditions and source waters were compiled to improve our understanding of the microorganism inactivation process using static mixers with ozone as a disinfectant. Data on transfer efficiency, microbial inactivation (utilizing *Giardia muris, Cryptosporidium parvum* and *Bacillus subtilis*), bromate formation, and destruction of micropollutants due to ozonation via a static mixer was gathered. The results were used to develop a hydrodynamic model (Chapter 5) and bromate formation model (Chapter 6) for static mixers. The conclusions specific to each phase of the study are presented in the individual Chapters of this report. The following is a listing of the major conclusions from each part of the research project and it also includes a synthesis of conclusions from the different Chapters as they relate to one another. The recommendations for future work are noted in the final section of this Chapter.

STATIC MIXERS FOR OZONE DISSOLUTION

Ozone dissolution performance of static mixers were studied at lab-, pilot- and full-scale. Based on the results obtained, the following conclusions can be drawn:

1. The ozone transfer efficiency is primarily influenced by static mixer hydrodynamic parameters i.e. superficial liquid velocity, v_s, and the gas-liquid flow rate ratio, G/L, within the static mixer. The ozone dose and the ozone residual at the outlet of the static mixer have little influence on the transfer efficiency compared to v_s or G/L.
2. Transfer efficiency was greatest at maximum v_s and minimum G/L. For the waters tested, a G/L ratio of 0.6 to 1.5 % and a liquid velocity of 0.7 to 1.6 m/sec resulted in the highest transfer efficiency (96% to 99%). Beyond this velocity, there was no further improvement of transfer efficiency. This velocity corresponded to a pressure drop of 10 to 20 KPa (1.5 to 2.9 psi).
3. A mathematical model was developed that related transfer efficiency to the superficial liquid velocity, v_s, and the gas-liquid flow rate ratio, *G/L*, within the static mixer. Modeling of the results showed that 94 % of the ozone transfer occurred in the static mixer itself and the gas-liquid separator provides little additional transfer. The static mixer was capable of achieving close to maximum theoretical ozone mass transfer.
4. Data collected from full-scale plants utilizing static mixers showed that transfer efficiencies of 95 to 98% were achieved during regular operations. Transfer efficiencies greater than 99% were recorded at the Clay Lane plant in the UK, which employs static mixers in the main line and sidestream.

INACTIVATION OF MICROORGANISMS AND BROMATE FORMATION

Inactivation of *Bacillus subtilis*, *Cryptosporidum parvum* and *Giardia muris* were studied at laboratory-scale. At pilot-scale, inactivation of *Bacillus subtilis* was studied. Inactivation of *Bacillus subtilis* spores, indigenous aerobic spores and coliforms were studied at full-scale. Bromate formation was studied at pilot- and full-scale. The following conclusions are drawn from these studies:

1. The residence time distribution of the lab-scale and pilot–scale units was characteristic of a plug flow with dispersion type hydraulic regime and no short-circuiting. This hydraulic characteristic was used to achieve efficient microbial inactivation in disinfection experiments. The lab-scale results showed that an integrated ozone disinfection process consisting of a static mixer, gas-liquid separator and a plug flow contactor can achieve microbial inactivation that is comparable to an idealized well-mixed batch reactor or plug flow reactor given the same average ozone concentration and contact time. Disinfection results obtained in the lab-scale contactor can be used to predict disinfection in the pilot-scale contactor. The disinfection process is scale-able. Kinetic results obtained from lab-scale reactors can be used to predict performance in larger continuous-flow ozone contactors.
2. No significant microbial inactivation was observed at the outlet of the static mixer. In challenge experiments with *Bacillus subtilis* spores, *Cryptosporidium parvum* oocysts and *Giardia muris* cysts at lab-scale, there was little evidence of significant microorganism inactivation due to direct gas-liquid ozone contact within the static mixer itself.
3. Although no microbial inactivation was measured within the static mixer itself, the static mixer provided the necessary starting conditions for efficient inactivation in the overall contacting system. These were (i) rapid and efficient dissolution of gaseous ozone and (ii) generation of a homogeneous dissolved ozone solution. The rapid and efficient dissolution of gaseous ozone lead to better Ct values thereby effectively exploiting the volume of the contactor. The generation of a homogeneous dissolved ozone solution was possible due to the micromixing environment offered by the static mixer. The lab study also demonstrated that a static mixer, when integrated into a well-designed, continuous-plug flow ozone contacting system, can achieve excellent and predictable *Cryptosporidium parvum* inactivation. In such a system, *Cryptosporidium parvum* inactivation can be predicted using relatively straightforward modeling techniques. Measured *Cryptosporidium parvum* inactivation was equivalent to (or slightly better) than that predicted using the published kinetic model of inactivation by ozone assuming plug flow hydrodynamics.
4. Results from pilot-scale studies conducted at the various sites showed that higher ozone residuals (compared to a FBD unit) were obtained in a short interval of time at the outlet of the static mixer due to intense mixing and ozone transfer through the static mixer elements. Enhanced microbial inactivation was observed due to a consequence of the higher Ct values reached along the static mixer unit.
5. A comparison of the results from pilot-scale studies conducted at the various sites in terms of ozone dissolution performance, microbial inactivation and bromate formation showed that water quality parameters (temperature and pH among other factors)

influence disinfection efficiency and bromate formation. The effect of temperature on inactivation of *Bacillus subtilis* with the pilot-scale unit confirmed the results obtained with the laboratory-scale unit and the well-stirred batch reactor. To achieve comparable inactivation, higher Ct values are required at cold temperatures. High disinfection efficiency and low bromate formation were observed at low pH (<7.0). The results underscored the importance of ozone dose, pH and alkalinity on bromate formation.

6. The Ct values for *Bacillus subtilis* and *Cryptosporidium parvum* obtained during laboratory-scale, pilot-scale studies, modeling and from literature were compared at different temperatures. The comparison indicated that at higher temperatures (≥ 16°C), the Ct requirements for similar levels of inactivation of both microorganisms are comparable. At temperatures 12°C to 3°C, *Bacillus subtilis* is inactivated more readily than *Cryptosporidium parvum*. The comparison of Ct values at these temperatures showed that 2-log inactivation of *Cryptosporidium parvum* can be achieved at 2-2.3 times the equivalent *Bacillus subtilis* Ct value. The modeling effort showed that Ct requirements for 1 log-unit inactivation of *Cryptosporidium parvum* or *Bacillus subtilis* are similar at all temperatures. On comparing Ct values for 2 log inactivation of *Cryptosporidium* (Finch, unpublished) at high and low temperatures with the Cts obtained from this study either thru lab work or modeling, it was found that the Ct required was reduced by 42-52%. Similar trends were seen while comparing Ct values derived in natural waters (Oppenheimer et al. 1999) and this work. This could have serious implications on the development of Ct tables for inactivation of *Cryptosporidium parvum*.

   The SWTR Guidance Manual provides technical guidance for the design of surface water treatment facilities (1990). The guidance manual focuses on the inactivation of *Giardia* and viruses, and Ct values for ozone inactivation of *Giardia* are provided in Appendix O. No Ct values are specified for the inactivation of *Cryptosporidium*; however, a commonly used method for determining *Cryptosporidium* inactivation has been to apply a linear multiplier to the tabulated *Giardia* Ct values to determine the *Cryptosporidium* Ct. In the absence of guidelines, ozone system designers have typically used multipliers ranging from 5 to 20 times *Giardia* Ct values to determine Cryptosporidium inactivation Ct values. In comparison, the potential for *Bacillus subtilis* to serve as a surrogate for *Cryptosporidium parvum* appears to be much better. Disinfection studies can be complex to design and construct, especially when pathogenic protozoa (i.e. *Cryptosporidium* and *Giardia*) are included. In such studies and other performance-based testing, where quantifying the inactivation of target microorganisms (e.g. *Cryptosporidium*) are required, *Bacillus subtilis* will serve as a very useful surrogate as they are easy to culture and enumerate. *Bacillus subtilis* spores are also useful for evaluation of the hydrodynamic efficiency of alternative ozone contactor designs, as they were found to be relatively resistant to ozone over a large temperature and pH range.

7. The results from full-scale studies showed that the inactivation of naturally occurring aerobic spores and *Bacillus subtilis* varied between 0.2 and 2.2 log. In general, *Bacillus subtilis* was less resistant to ozone than the other microorganisms tested. An optimal ozone dose of 0.2 to 0.4 mg O_3/mg C ensured bromate below the MCL

(10 ppb) at all times even though the pH varied between 7.2 and 8.3. Bromate formation was shown to be influenced by the contact time, ozone dose, alkalinity, pH and the level of bromide in the inlet water. pH was the most influential parameter that controlled bromate formation.

8. In general, applying low ozone doses (≤ 0.5 mg O_3/mg C) through a static mixer ozone dissolution system integrated with a well designed downstream contactor (plug flow) combined with a pH (≤ 7.0) can provide adequate microbial inactivation while keeping bromate formation below the current MCL (10 ppb). For some waters, it may be necessary to depress the pH to 6.5 to achieve similar results. The most challenging waters to treat will be extremely cold waters with significant concentrations of bromide.

STATIC MIXERS VERSUS FINE BUBBLE DIFFUSION CONTACTORS

Pilot studies were conducted at the Los Angeles Department of Water and Power, CA (LADWP) and Elizabethtown Water Company, NJ (EWC) where the static mixer unit (SMU) was compared side-by-side with a FBD unit. Data from a full-scale plant in the UK (Iver) which employed FBD contactors and turbine diffusers was also garnered to compare their effectiveness against plants that employed static mixers for ozone dissolution. The following conclusions are made based on the results of these studies:

1. The results of this study showed that intense mixing and ozone transfer through the static mixer elements leads to a much faster ozone dissolution and systematically higher Cts along the SMU (1.2-2.3 times) than those provided through the FBD unit of equivalent contact time. Inactivation data obtained with *Bacillus subtilis* at LADWP and EWC showed that the SMU provided enhanced inactivation (0.4 to 2.9 Log more than FBD unit) under similar applied ozone doses, water quality conditions, and equivalent contact times. This improvement was due to higher Ct values reached along the SMU. While utilizing the SMU at the LADWP, higher microbial inactivation was achieved at an adjusted pH of 6.5 than the natural water pH due to a higher Ct obtained at the lower pH. Also, bromate formation was minimized at the lower pH (6.5). For cold waters, such as the one encountered in EWC (5°C), higher Ct was necessary to achieve comparable levels of microbial inactivation. No significant difference in bromate formation was observed between SMU and FBD units while testing them at LADWP and EWC.
2. The results from full-scale studies showed that static mixers are more effective for inactivation of microorganisms than conventional diffusers, particularly for the inactivation of coliforms. The inactivation at the Iver plant was significantly lower as compared to plants that employ static mixers. Also, the lowest inactivation of aerobic spores and *Bacillus subtilis* spores was observed at Iver.

HYDRODYNAMIC AND BROMATE FORMATION MODELING

Hydrodynamic modeling was performed to model the ozone transfer and consumption through the static mixer and the gas-liquid separator utilizing bench and pilot data incorporating natural water quality conditions. A satisfactory simulation was obtained for the ozone concentration (residual) at the outlet of the static mixer or of the gas/liquid separator, ozone consumption rate and transfer efficiency. The model simulated the transfer and reactivity of ozone in static mixers for various configurations, as well as the contribution of the gas/liquid separator quite well.

Bromate formation models were developed incorporating the following parameters: bromide concentration, ozone dose, pH, temperature, DOC, contact time and alkalinity. The simulations for the SMU showed that predicted bromate was in agreement with measured bromate.

ADVANTAGES OF STATIC MIXERS FOR OZONATION

Vivendi Water operates numerous full-scale plants in France and the UK equipped with static mixers for pre-ozonation and intermediate ozonation and has extensive experience in their operation. Table B.4 of Appendix B shows several full-scale plants employing static mixers for pre- and intermediate ozonation. Based on the data gathered at bench-, pilot- and full-scale plants from this project and the operational experience of full-scale plants employing static mixers in Europe the advantages of using static mixers for ozonation can be summarized as follows:

1. Static mixers exhibit very high transfer efficiencies (up to 99% were recorded in tests in the UK full-scale plants; Davidson, 1997), are very effective in destroying pesticides such as atrazine, inactivating microorganisms, removal of color and eliminating taste and odor problems, all within a few seconds. The improvement in transfer efficiencies leads to lower applied ozone dosages to achieve equivalent Ct conditions, leading to lower operational costs.
2. In a static mixer, the mass transfer coefficient (k_La) ranges from 0.1 to 10 s^{-1}. Transfer coefficient of about 5 s^{-1} for 400 W/kg of dissipated power have been reported (Middleton, 1978), which is 100 to 1000 times greater than for a baffled reactor or a classical gas/liquid separator (Hérès, 1993). The results from this study suggested that for a given ozone concentration, the most efficient operation is at a low G/L and high feed gas ozone concentration. Improved transfer efficiency at low gas flows is an advantage of static mixers because it means that these devices can be effectively integrated with modern, high-concentration, medium frequency ozone generators. In comparison, conventional fine bubble diffusion systems are limited by a requirement of sufficient gas flow rate to ensure adequate gas-liquid mixing and ozone dissolution (Schulz et al. 1995).
3. The results from this study have important implications for the design of ozone contactors with static mixers in full-scale plants. The FBD systems require large ozone injection/transfer chambers for ozone dissolution. The results of this study showed that intense mixing and ozone transfer through the static mixer elements leads to a much faster ozone dissolution and systematically higher Ct values along the SMU than those provided through the FBD unit. This study also showed that Ct requirements could be lowered by 42% to 52% to achieve the same level of Cryptosporidium inactivation. The utilization of static mixers as an ozone dissolution technology can lead to reduction in the contactor size downstream of the ozone dissolution system as targeted inactivation efficiencies are reached more quickly with an optimized Ct as compared to FBD systems. As the hydraulic residence time (HRT) of the static mixer is very small (few seconds), the overall footprint of the ozonation process is also significantly

reduced when compared to a FBD system which needs a large injection chamber for ozone dissolution. This overall reduction in HRT (25 to 45% reductions were seen in this study) for the ozone contactors results in significant savings in capital, operation and maintenance (O & M) costs.

4. Static mixer systems produce a homogeneous ozone solution, have no moving parts, are easy to retrofit, require minimal maintenance, have low capital and O & M costs, have low energy consumption, possess wide turndown ratios, are adaptable to automatic process control, are simple to operate, and have no small holes which could be gradually plugged. Headlosses in the static mixer system can be significantly reduced by adopting a sidestream option to a foot or less (see Chapter 4 and Figure B.1 of Appendix B). The porous plates of FBD systems have small holes which get plugged (iron or/and manganese) requiring frequent maintenance, the porous plates themselves have to be replaced every 5 years and the rubber gaskets are subject to fouling/wear and tear. Table B.5 (Appendix B) presents a cost comparison for a typical drinking WTP with the static mixer versus FBD system option for ozone dissolution.

RECOMMENDATIONS FOR FUTURE WORK

The work done in this project helped identify several important areas for future research that are of value to the drinking water industry. Specific recommendations are listed below:

1. Inactivation of *Cryptosporidium parvum* oocysts by ozone were conducted with the laboratory study static mixer ozone contactor at the feed water temperature of 22°C. More inactivation data needs to be developed at cold water temperatures especially in continuous flow systems such as the one used in this study. This data needs to be compared to inactivation data obtained for *Bacillus subtilis* (also at cold temperatures) to develop the equivalent Ct value for inactivation of *Cryptosporidium parvum*. These experiments are necessary as spore inactivation characterisitics are sensitive to artifacts such as clumping and agglomeration and may vary from batch to batch. This would help in further confirming the suitability of *Bacillus subtilis* as a surrogate to *Cryptosporidium parvum*. Also future research should be done using oocysts that have been exposed to environmental conditions that are representative of a real world system, rather than using "fresh" laboratory preserved oocysts.
2. The bench and pilot studies should compare the performance of different types of static mixers. The effects of other static mixer design parameters like the number of mixing elements and the diameter of the static mixer need to be studied.
3. More tests need to be done to study the effect of ammonia in lowering bromate formation. Future experimental work should investigate different injection points for introducing ozone in the static mixer as some studies have shown that this results in lowered bromate formation.
4. Although the hydrodynamic model used in this work resulted in a good agreement with the available data, the use of adjusted parameters should be reduced to the minimum for the design of ozone transfer. Parameters for modeling bromate formation need to be included in this model.

5. The *Ozone System Energy Optimization Handbook* (DeMers et al. 1996) includes only an investigation of FBD contactors. The objective of that study was to develop and demonstrate energy savings opportunities while achieving water quality goals at water treatment plants using ozone. Use of a static mixer system is another step in the optimization of energy for ozone applications by increasing ozone transfer efficiency, optimizing Ct and thereby reducing capital and O & M costs. These benefits need to be incorporated in the *Handbook* and in AwwaRF's Integrated Disinfection Framework Project. This should provide a good scientific basis for utilities considering to implement ozonation at their treatment plants.

APPENDIX A
LAB-SCALE EXPERIMENTAL DATA

Table A.1
Laboratory study batch reactor ozonation trials with *Cryptosporidium parvum* oocysts

				Direct UV Absorbance @ 260 nm						Indigo Trisulphonate Colorimetric Method						
Trial*	pH	C_a mg/L	t_b min	C_0 mg/L	C_f mg/L	$C_a - C_0$ mg/L	k_d /min	C_{avg} mg/L	$C_{avg}t$ mg × min/L	C_0 mg/L	C_f	$C_a - C_0$ mg/L	k_d /min	C_{avg} mg/L	$C_{avg}t$ mg × min/L	$-\log(N/N_0)$
1	7.8	1.43	12	1.55	0.23	-0.12	0.16	0.69	8.34	1.72	0.31	-0.29	0.14	0.82	9.85	2.7
2	7.9	0.8	12.3	0.88	0.2	-0.08	0.12	0.46	5.63	0.95	0.18	-0.15	0.13	0.46	5.7	2.8
3	7.9	0.79	4	1.04	0.5	-0.25	0.18	0.74	2.94	0.99	0.23	-0.2	0.13	0.78	3.13	1.9
4	7.8	1.35	12	1.43	0.24	-0.08	0.15	0.67	8.02	1.79	0.3	-0.44	0.15	0.83	9.98	2.8
5	7.8	0.79	4.9	0.77	0.31	0.02	0.19	0.51	2.46	0.85	0.4	-0.06	0.16	0.6	2.9	1.4
6	7.9	0.79	4	0.81	0.28	-0.02	0.26	0.5	1.99	1.00	0.33	-0.21	0.28	0.61	2.42	2.2
7	7.9	0.79	12	0.77	0.04	0.02	0.25	0.24	2.91	0.92	0.04	-0.13	0.27	0.27	3.28	2.3

* All trials were conducted in samples of feed water from the static mixer ozone contactor at 22°C

Table A.2
Information from bath ozonation trials with *Giardia muris* cysts extracted from Charles Labatiuk Ph. D. thesis (1992) and simplified Hom model predictions

Trial*	pH	Batch Time t_b min	Applied Ozone C_a mg/L	Initial Ozone C_0 mg/L	Final Ozone C_f mg/L	C_{avg} mg/L	$C_{avg}t_b$ mg × min/L	Measured -log N/N_0	Hom Model Prediction -log N/N_0	Model Error
A1	6.7	5	0.62	0.47	0.3	0.38	1.88	4.2	3.71	0.49
A2	6.7	5	0.64	0.52	0.4	0.46	2.28	4.2	3.84	0.36
B1	6.7	5	0.26	0.18	0.1	0.13	0.67	3.1	3.10	0.00
B2	6.7	5	0.3	0.22	0.1	0.15	0.74	3.1	3.15	-0.05
C1	6.7	5	2.55	2.14	1.3	1.67	8.34	4.4	4.83	-0.43
C2	6.7	0.58	0.13	0.07	0.04	0.05	0.03	1.2	1.79	-0.59
D1	6.7	0.58	0.28	0.15	0.2	0.17	0.10	2.2	2.21	-0.01
30-4	7.6	2	2.73	2.1	1.36	1.69	3.38	3.4	4.12	-0.72
30-7	7.6	7	1.54	1.48	0.29	0.66	4.59	4	4.35	-0.35
36-4	7.6	5	0.82	0.7	0.36	0.50	2.51	5.1	3.91	1.19

* All trials were conducted in ozone demand-free 0.05 M phosphate buffered ultrapure water.

APPENDIX B
FULL-SCALE STATIC MIXER INSTALLATIONS IN THE UK

DETAILS OF PROCESS TRAINS AT UK WTPS

Chertsey WTW	*Stream 1:* Groundwater (river gravels) Pre-ozonation Rapid gravity filtration *Stream 2:* Surface water - bankside storage Rapid gravity filters Slow sand filtration *Combined streams:* Inter-ozonation GAC filtration Disinfection (super- and de-chlorination)
Clay Lane WTW	Groundwater source abstracted from eight remote borehole sites Single stage ozonation GAC filtration Disinfection (super- and de-chlorination)
Egham WTW	Surface water - direct river abstraction Pre-ozonation Coagulation/clarification Rapid gravity filtration Inter-ozonation GAC filtration Disinfection (super- and de-chlorination)
Iver WTW	Surface water - direct river abstraction Pre-ozonation Coagulation/clarification Inter-ozonation GAC filtration Disinfection (super- and de-chlorination)
Walton WTW	Surface water - river with bankside storage Pre-ozonation Rapid gravity filtration Slow sand filtration Inter-ozonation GAC filtration Disinfection (super- and de-chlorination)

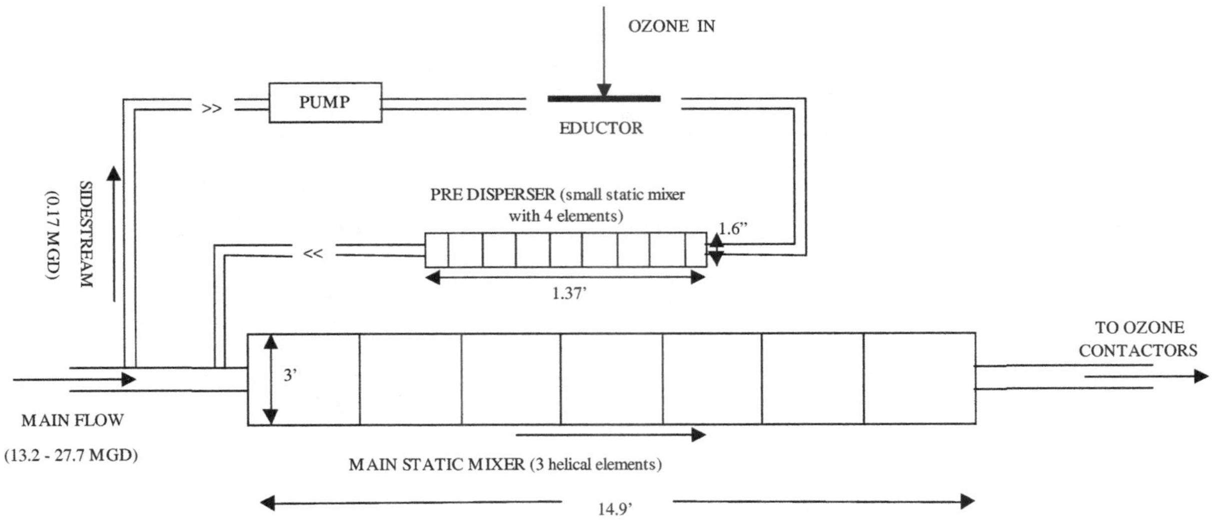

Figure B.1 Configuration of the ozone dissolution system at the Clay Lane WTP, UK

Table B.1
Static mixer configurations

Site	Walton	Walton	Chertsey	Chertsey	Egham	Egham	Clay Lane
Duty	Pre – ozonation	Inter - ozonation	Pre - ozonation	Inter - ozonation	Pre - ozonation	Inter - ozonation	Inter - ozonation
Configuration	Horizontal	Horizontal	Horizontal	Vertical Down	Horizontal	Horizontal	Horizontal
Type of Ozone Application	N.A	Direct Injection	N.A	Direct Injection	N.A	Direct Injection	Direct and Side – stream Injection
Liquid Flowrate (m^3/h)	0.68 to 1.93	0.25 to 1.25	0.91 to 2.52	0.91 to 1.95	4.2 to 9.0	2.18 to 5.22	3.61 to 6.29
Gas Flowrate (scfm)	15.3	7.1 – 17.1	N.A	N.A	17.2 – 73.6	15.9 – 74.7	86.5 – 151.9
Ozone Dose (mg/L)	2.0	4.0	1.5 – 3.0	1.5 – 4.0	2.0	3.5	5.9
Mixer Diameter (mm)	600	500	600	N.A	1,000	700	900
Mixer Specificaton	N.A	4 elements, No empty elements	N.A	N.A	N.A	4 elements, 2SM-1E-1SM-1E-1SM	4 elements, 2SM-1E-1SM-1E-1SM
Total Mixer Length (m)	N.A	1.6	N.A	N.A	N.A	3.9	4.5
Pressure Drop (psi)	N.A	2.6 @ 1.7 m^3/h	N.A	N.A	N.A	0.7 to 3.8	1.2 to 3.3

Where: SM = Mixing element
E = Empty space

Each unit is implemented with SMV static mixers from Sulzer Chemtech.

Table B.2
Ozone contactor configurations

Site	Walton	Walton	Chertsey	Chertsey	Egham	Egham	Clay Lane
Duty	Pre - ozonation	Inter - ozonation	Pre - ozonation	Inter - ozonation	Pre - ozonation	Inter - ozonation	Inter - ozonation
Contact Tank Volume (m^3)	22.0	63.2	33.8	84.3	105	333.4	324.0
Contact Time (min)	1.0 to 2.5	3.2 to 7.6	1.1 to 3.2	3.4 to 8.0	N.A	5.3 to 13.1	6.3 to 7.5
Depth (m)	5.5	5.5	6.5	9.7	2.2	5.7	6.0
Width (m)	2.0	5.0	2.0	3.0	3.8	6.6	6.0
Number of Compartments	1	4	1	1	1	4	4
Flow Direction	↑	↑↓↑↓	↑	↑	↑	↑↓↑↓	↑↓↑↓
Length of each Compartment (m)	2.0	1.5/ 2.4/ 2.4/ 1.5	2.6	3.0	8.6	2.0/ 3.5/ 3.6/ 1.5	3.0/ 3.0/ 3.0/ 1.6

Where ↑ = Upward flow
↓ = Downward Flow

Table B.3
Raw water quality at the WTP's in UK

WTP	Alkalinity mg/L as $CaCO_3$	Atrazine µg/L	Bromide µg/L	pH	Temp °C	TOC mg/L as C	Turbidity NTU
Chertsey							
River Thames	242-289	0.02-0.14	72-171	7.8-9.0	3-18	2.9-6.4	0.4-94.5
Abbeymeads	234-307	0.01-0.07	115-281	7.5	8-9	2.5-3.7	0.1-0.6
Clay Lane	327-337	0.01-0.21	90-209	6.8-7.3	9-11	1.3-4.1	0.1-2.4
Egham	256-289	0.01-0.12	75-164	7.4-7.9	4-22	2.7-8.3	0.3-115
Iver	212-305	0.01-0.15	73-133	7.7-8.7	11-19	3.4-9.3	0.2-64
Walton	176-280	0.01-0.21	66-157	7.9-9.1	11-18	4.0-6.4	1.0-78.1

Table B.4
Full-scale plants employing static mixers for ozonation

Site	Country	Location of Ozone Treatment	Water Flow (m^3/d)	Number of parallel streams	Side-stream with static mixer	Configuration	Mixer diameter/ number of mixing elements
Walton*	UK	Pre- and Intermediate Ozonation	57,000	2,2	No	Horizontal Horizontal	1.7'/4
Chertsey*	UK	Same as above	57,000	3,2	No	Horizontal Vertical	2'/4, 1.64'/
Egham*	UK	Same as above	151,000	3,2	No	Horizontal Horizonal	3.3'/4, 2.3'/
Tophill Low*	UK	Pre-Ozonation	72,000	1	Yes	Horizontal	
Clay Lane*	UK	Pre-Ozonation	151,000	2	Yes	Horizontal	3.0'/4
Melunada Arvigny*	France	Intermediate	83,000	1	Yes	Vertical	2.6'/4
Agen	France	Pre-ozonation	11,300	1	No	Horizontal	
Bouillon-la-Graviere*	France	Post-Ozonation	11,300	1	No	Horizontal	1.64'
Ardleigh*	UK	Intermediate	98,000	2	No	Horizontal	
CSK Mol*	Belgium	Pre-ozonation	500	1	No	Horizontal	0.33'
Mery-sur-Oise*	France	Intermediate	170,000	4	Yes	Vertical	3.3'
St-Chamond*	France	Pre-Ozonation	15,000	1	No	Horizontal	
Leon*	Italy	Pre-Ozonation	2,600	1		Horizontal	
Tampa Bay†	USA	Intermediate	250,000	2	Yes	Vertical	3.3/3
Wiggins WTP	South Africa	Pre-Ozonation	350,000	4	No	Horizontal	3.3/5

* Operated by Vivendi Water-Europe

† Under construction and to be operated by US Filter/Vivendi Water-USA.

wide turndown (6:1), headloss through this system is less than 0.3 m (1 foot)

Table B.5
Cost comparison for a typical WTP

	Alternative A with static mixer & sidestream	Alternative B with FBD	Comments
Number of Static Mixers/ Injection Chambers	2	4	4 injection chambers (total) for the FBD system, no. of reaction chambers same for both systems
Capital Cost ($)	60,000	250,000	Injection chamber cost includes cost of land ($100/sq. ft), concrete, steel piping, diffusers, and diffuser manifold.
Capital Cost of standby Pump ($)	3,000		
Consumed Energy in KW	2.2 (1.1 KW/train)		
Present Worth Energy Cost ($)	14,000		I=6%, N=20 years,$0.06/KW-hr
Spare Parts (approximate)	10,000	70,000	• Pumps every 10 years for the GDS system • Diffusers every 5 years, diffuser manifold every 15 years for the FBD system
Total Present Worth Cost for a 20-year operation ($)	87,000	320,000	

The cost of a static mixer system with a side stream option (Figure B.1) versus the fine bubble diffusion (FBD) contactor for treating 15 MGD flow (two trains) is given above. The cost comparison is only between the dissolution system used for both systems, i.e the static mixer and the injection chambers of the FBD system.

Notes:

(i) Add to these costs the cost of piping, ozone generation equipment, chemicals, compressors, ozone destruction unit and ancillary equipment which should be similar for the static mixer or the FBD system. *The cost of the reaction chambers for the static mixer system will be lower than the FBD system as the overall HRT is 25% to 45% less for static mixer system to achieve similar Ct values.*

(ii) The costs for the static mixer with sidestream option were obtained from Statiflo, UK and the costs for the FBD system was obtained from Langlais et al. (1991).

APPENDIX C
MASS TRANSFER COEFFICIENTS

Table C.1
Correlation for the estimation of the overall mass transfer coefficient

Authors	Correlation	Studied system	Operating conditions
Middleton, 1978	$K_l a = 1.74\ 10^{-4}\left[\dfrac{Puis}{V_d}\right]^{0.8}$	O_2 (air) - Eau	*Static mixers :* Sülzer, Kenics *Configuration :* Vertical downwards U_{SL} = 0.2 to 1.0 $m.s^{-1}$ – G/L < 6
Zhou, 1991	$K_l a = 4.56\ 10^{-4}\left[\dfrac{Puis}{V_d}\right]^{0.68}$	O_2 (air) - Eau	*Static mixers :* Kenics *Configuration :* Horizontal U_{SL} = 0.6 to 3.2 $m.s^{-1}$
Roes et al., 1984	$K_l a = 0.63\left[\dfrac{Puis}{M}\right]^{0.75}.\varepsilon_g$	O_2 (air) - Eau	*Static mixers :* Sülzer SMV *Configuration:* Vertical upwards U_{SL} = 0.2 to 1.0 $m.s^{-1}$ / G/L < 1 Bubble regime
Hérès, 1993	$K_l a = 2.01\ 10^{-3}\left[\dfrac{Puis}{V_d}\right]^{0.65}.\varepsilon_g^{\ 0.7}$	O_3 (O_2) - Eau	*Static mixers :* Sülzer SMV bubble, intermediate and annular regime
Martin, 1992	$K_l a = C^{ste}\left[\dfrac{Puis}{V_d}\right]^{\beta}\varepsilon_g^{\ \gamma}$		*Configuration:* Vertical upwards
	or $K_l a = C^{ste}.L^{(2.\beta-\gamma)}.G^{\gamma}$ $\beta = 0.7 \pm 0.1$ $\gamma = 0.8 \pm 0.1$	O_3 (O_2) - Eau	*Static mixers :* Sülzer SMV
	$\beta = 0.5 - 0.55$ $\gamma = 0.6 - 0.75$	O_3 (Air) - Eau	*Static mixers :* Statline
Zhou, 1992	$K_l a = 3.76\ 10^{-3}\left[\dfrac{Puis}{V_d}\right]^{0.6}.\varepsilon_g^{\ 0.6}.(1-\varepsilon_g)$	O_3 (Air) - Eau	*Static mixers :* Sülzer SMV, Kenics *Configuration :* Vertical downwards U_{SL} = 0.2 to 2.6 $m.s^{-1}$ bubble, intermediate and annular regime
Grosz-Röll et al., 1983	$K_l a = 7.2\ 10^{-3}\left[\dfrac{Puis}{V_d}\right]^{0.69}.\varepsilon_g^{\ 0.8}.d_h^{0.26}$	O_3 (O_2) - Eau	*Static mixers :* Sülzer SMV *Configuration:* Horizontal U_{SL} = 0.3 to 1.0 $m.s^{-1}$ / G/L < 0,3

REFERENCES

Ahmad, M., and S. Farooq. 1984. Influence of Bubble Sizes on Ozone Solubility Utilization and Disinfection. *Water Science and Technology*, 17:1081-1090.

Amitharajah, A., S. Casey Jones, B.M. Skeens, H.L. Heindel, W. Li, S.A. Hardy, and R. Latimer. 2001. Static Mixer for Coagulation and Disinfection. AwwaRF Report.

APHA AWWA WEF. 1992. *Standard Methods for the Examination of Water and Wastewater*. 18th edition. Washington, DC, USA: American Public Health Association.

AWWA and AwwaRF. 1992. *Water Industry Data Base*. Denver, CO: AWWA and AwwaRF.

Bader, H., and J. Hoigné. 1981. Determination of Ozone in Water by the Indigo Method. *Water Research*, 15:449-456.

Bader, H., and J. Hoigné. 1982. Determination of Ozone in Water by the Indigo Method: A Submitted Standard Method. *Ozone: Science and Engineering*, 4:169-176.

Bais, N. 2000. Modélisation des Mélangeurs Statiques en tant que Contacteurs d'Ozone en Présence de Matières Organiques Naturelles. DEA report, University of Rennes and University of Poitiers, France.

Barbeau, B., L. Boulos, R. Desjardins, J. Coallier, M. Prévost, M., and D. Duchesne. 1997. A Modified Method for the Enumeration of Aerobic Spore-Forming Bacteria. *Canadian Journal of Microbiology*, 43(10):976-980.

Barbeau, B., L. Boulos, R. Desjardins, J. Coallier, and M. Prévost. 1999. Examining the Use of Aerobic Spore-Forming Bacteria to Assess the Efficiency of Chlorination. *Water Research*, 33(13): 2941-2948.

Bellamy, W., K. Carlson, D. Pier, J. Ducoste, and M. Carlson. 2000. Determining Disinfection Needs. *Journal of the American Water Works Association*, 92(5):44-52.

Belosevic, M.G., and E. Faubert. 1983. *Giardia muris*: Correlation Between Oral Dosage, Course of Infection, and Trophozoite Distribution in the Mouse Small Intestine. *Laboratory Parasitology*, 56:93-100.

Belosevic, M., G. Faubert, E. Skamene, and J.D. MacLean. 1984. Susceptibility and Resistance of Inbred Mice to *Giardia muris*. *Infection and Immunity*, 44:282-286.

Bin, A.K., and M. Roustan. 2000. Mass Transfer in Ozone Reactors. IOA symposium "Fundamental and Engineering Concepts for Ozone Reactor Design." Toulouse, France, March 1-3.

Bonnard, R. 1998. Optimisation de la Mise en Oeuvre des Mélangeurs Statiques comme Moyen de Transfert lors d'une Etape d'Ozonation en Traitement d'Eau Potable. Mémoire de fin d'étude. CNAM Paris.

Bonnard, R., C. Adrien, C. de Traversay, and F. Luck. 1999. Static Mixer as a Disinfectant Process? In *Proceedings of the 14th Ozone World Congress, August 22-26, 1999*. Volume 2, pp. 251-261. Dearborn, MI: International Ozone Association.

Botzenhart, K., G. Tarcson, and M. Ostruschka. 1993. Inactivation of Bacteria and Coliphages by Ozone and Chlorine Dioxide in a Continuous Flow Reactor. *Water Science and Technology*, 27(3-4):363-370.

Bourne, J.R., and H. Maire. 1991. Micromixing and Fast Chemical Reactions in Static Mixers. *Chem. Eng. Process*, 30(1):23-30.

Bourne, J.R., J. Lenzner, and S. Petrozzi. 1992. Micromixing in Static Mixers. An Laboratory Study. *Ind. Eng. Chem. Res.*, 31(4):1216-1222.

Box, G.P., W.G. Hunter, and J.S. Hunter. 1978. *Statistics for Experimenters*. John Wiley & Sons, New York, NY.

Brand, R.J., D.E. Pinnock, and K.L. Jackson. 1973. Large Sample Confidence Bands for the Logistic Response Curve and its Inverse. *The American Statistician*, 27(4):157-160.

Broadwater, W.T., R.C. Hoehn, and P.H. KinGiardia 1973. Sensitivity of Three Selected Bacterial Species to Ozone. *Applied Microbiology*, 26(3):391-393.

Calderbank, P.H. 1967. Gas Absorption from Bubbles. *Chem. Eng.*, 45, 209.

Carlson, K., W. Bellamy, J. Ducoste, and G. Amy. 2001. Implementation of the Integrated Disinfection Design Framework, AwwaRF Report.

Cerf, O. 1977. A Review: Tailing of Survival Curves of Bacterial Spores. *Journal of Applied Bacteriology*, 42:1-19.

Chandra, G.K., and D.D. Kale. 1992. Pressure Drop for Laminar Flow of Viscoelastic Fluids in Static Mixers. *Chem. Eng. Sci.*, 47 (8):2097-2100.

Charpentier, J.C., and G. Wild. 1983. *Absorption avec Réaction Chimique*. Ed. Technique de l'Ingénieur.

Choe, K. 1998. Ozone Inactivation of *Bacillus cereus* Spores in Ozone Demand Free Water. Master of Science Thesis, University of Alberta, Edmonton, Alberta.

Coulson, J.M., and J.F. Richardson. 1993. *Chemical Engineering, Vol. I: Fluid Flow, Heat and Mass Transfer*. 4th ed., Lavoisier, Paris.

Cybulski A., K. Werner. 1986. Static Mixers: Criteria for Application and Selection. *Inst. Chem. Eng.*, 26 (1), 171-180.

Dahi, E., and E. Lund. 1980. Steady State Disinfection of Water by Ozone and Sonozone. *Ozone: Science and Engineering*, 2:13-23.

Danckwerts, P.V. 1951. Significance of Liquid-Film Coefficients in Gas Absorption. *Ind. Eng. Chem.*, 43, 1460-1467.

Danckwerts, P.V. 1955. Gas Absorption Accompained by Chemical Reaction. *AIChE J.*, 1, 456-463.

Davidson, C. 1997. Personal communication. Vivendi Water, UK.

DeMers, L.D., K.L. Rakness, and B.D Blank. 1996. Ozone System Energy Optimization Handbook. AwwaRF - EPRI Collaborative Research.

Do-Quang, Z., A. Cockx, and A. Line. 2000. Recent Advances in Modelling Tools Development and Application for Ozone Reactor Design: the CFD Approach. IOA symposium "Fundamental and Engineering Concepts for Ozone Reactor Design." Toulouse, France, March 1-3.

Doré, M. 1989. Chimie des Oxydants et Traitement des Eaux. Edition technique et documentation, Lavoisier, Paris, France.

Driedger, A., E. Staub, U. Pinkernell, B. Marinas, W. Koster, and U. Von Gunten. 2001. Inactivation of *Bacillus Subtilis* Spores and Formation of Bromate During Ozonation. *Water Research*. 35 (12), 2950-2960.

Dudley, J. 1995. Mass Transfer in Bubble Columns: a Comparison of Correlations. *Water Research*, 29 (4), 1129-1138.

Duguet, J.P. 2000. Ozone Demand of Waters. IOA Symposium "Fundamental and Engineering Concepts for Ozone Reactor Design." Toulouse, France, March 1-3.

Facile, N., B. Barbeau, M. Prevost, and B. Koudjonou. 2000. Evaluating Bacterial Aerobic Spores as a Surrogate for *Giardia* and *Cryptosporidium* Inactivation by Ozone. *Water Research*, 34(12):3238-3246.

Farooq, S., E.S.K. Chian, and R.S. Engelbrecht. 1977. Basic Concepts in Disinfection With Ozone. *Journal of the Water Pollution Control Federation*, 49:1818-1831.

Farooq, S., R.S. Engelbrecht, and E.S.K. Chian. 1978. Process Considerations in Design of Ozone Contactor for Disinfection. *Journal of the Environmental Engineering Division, Proceedings of the American Society of Civil Engineering*, 104(EE5):835-847.

Finch, G.R. 1999. Unpublished data presented to the USEPA M/DBP Technical Workshop, October 1999.

Finch, G.R., E.K. Black, and L.L. Gyürék. 1995. Ozone and Chlorine Inactivation of *Cryptosporidium*. In *Proceedings of the American Water Works Association Water Quality Technology Conference, November 6-10, 1994, San Francisco, CA*. pp. 1303-1318. Denver, CO: American Water Works Association.

Finch, G.R., E.K. Black, L. Gyürék, and M. Belosevic 1994. *Ozone Disinfection of Giardia and Cryptosporidium*. Awwa Research Foundation and American Water Works Association, Denver, CO.

Finch, G.R., E.K. Black, C.W. Labatiuk, L. Gyürék, and M. Belosevic 1993a. Comparison of *Giardia lamblia* and *Giardia muris* Cyst Inactivation by Ozone. *Applied and Environmental Microbiology*, 59(11):3674-3680.

Finch, G.R., and K. Choe. 1999. Using *Bacillus* Spores as a Surrogate for Ozone Inactivation of *Cryptosporidium*. In *Proceedings of the 14th Ozone World Congress, August 22-26, 1999*. Volume 1, pp. 35-42. Dearborn, MI: Interational Ozone Association.

Finch, G.R., C.W. Daniels, E.K. Black, F.W. Schaefer III, and M. Belosevic. 1993b. Dose-Response of *Cryptosporidium parvum* in Outbred, Neonatal CD-1 Mice. *Applied and Environmental Microbiology*, 59(11):3661-3665.

Finch, G.R., L.L. Gyürék, L.R.J. Liyanage, and M. Belosevic 1997. *Effect of Various Disinfection Methods on the Inactivation of Cryptosporidium*. Awwa Research Foundation and American Water Works Association, Denver, CO.

Finch, G.R., C.W. Labatiuk, R.D. Helmer, and M. Belosevic 1992. *Ozone and Ozone-Peroxide Disinfection of Giardia and Viruses*. Awwa Research Foundation and American Water Works Association, Denver, CO.

Finch, G.R., and D.W. Smith. 1991. Pilot-Scale Evaluation of the Effects of Mixing on Ozone Disinfection of *Escherichia coli* in a Semi-Batch Stirred Tank Reactor. *Ozone: Science and Engineering*, 13(5):593-605.

Fuchun, X., and L. Cunli. 1990. Mass Balance Analysis of Ozone in a Conventional Bubble Column. *Ozone: Science and Engineering*, 12 (3):269-279.

Glover, J.B., M.D. Scharbow, and J.B. Neethling. 1993. Direct Injection In-Line Contactor Design. *Proceedings of the IOA Conference*, Los Angeles, S14, 101-113.

Goepfert, J.M., W.M. Spira, and H.U. Kim. 1972. *Bacillus cereus*: Food Poisoning Organism. A Review. *J. Milk Food Technol.*, 35(4):213-227.

Gordon, G. 1987. The Very Slow Decomposition of Aqueous Ozone in Highly Basic Solutions. *Proceedings of the 8th Ozone World Congress*, IOA, Zurich, Switzerland.

Gordon, G., R.D. Guaw, M. Yuko, B. Walters, and B. Bubnis. 2000. Using Indigo Absorbance to Calculate the Indigo Sensitivity Coefficient. *Journal of the American Water Works Association*, 92(12):96-107.

Gordon, G., G.E. Pacey, W.J. Cooper, and R.G. Rice. 1988. Current State-of-the-Art Measurements of Ozone in the Gas Phase and in Solution. *Ozone: Science and Engineering*, 10:353-366.

Grosz-Röll, F., J. Bättig, and F. Moser. 1982. Gas/Liquid Mass Transfer With Static Mixer Units. In *Fourth European Conference on Mixing*. pp. 225-236. Noordwijkerhout, The Netherlands: BHRA Fluid Engineering.

Grozs-Röll, F., G.J. Batti, and F. Moser. 1983. Gas/Flussig Stoffubergang in Statischen Mischern. *Verfahrenstechnick*, 17 (12): 698-707.

Guidance Manual for Compliance With the Filtration and Disinfection Requirements for Public Water Systems Using Surface Water Sources, USEPA, October 1990.

Gyürék, L.L., and G.R. Finch. 1998. Modeling Water Treatment Chemical Disinfection Kinetics. *Journal of Environmental Engineering*, 124(9):783-793.

Gyürék, L.L., H. Li, M. Belosevic, and G.R. Finch. 1999. Ozone Inactivation Kinetics of *Cryptosporidium parvum* in Phosphate Buffer. *Journal of Environmental Engineering*, 125(10):913-924.

Gyürék, L.L., G.R. Finch, and M. Belosevic. 1997. Modeling Chlorine Inactivation Requirements of *Cryptosporidium parvum* Oocysts. *Journal of Environmental Engineering*, 123(9):865-875.

Haas, C.N. 1988. Micromixing and Dispersion in Chlorine Contact Chambers. *Environmental Technology Letters*, 9:35-44.

Haas, C.N. 1990. Disinfection. In *Water Quality and Treatment*. Edited by F. Pontius. pp. 877-932. McGraw-Hill, New York, NY.

Haas, C.N., and J. Joffe. 1994. Disinfection Under Dynamic Conditions - Modification of Hom Model for Decay. *Environmental Science and Technology*, 28(7):1367-1369.

Haas, C.N., M. Heath, J. Jacangelo, U. Anmangandla, J. Joffe, J.C. Hornberger, and J. Glicker. 1994. *Development and Validation of Rational Design Methods of Disinfection.* Denver, CO: AWWA and AwwaRF.

Haas, C.N., J. Joffe, U. Anmangandla, G. Jacangelo, and M. Heath. 1996. Water Quality and Disinfection Kinetics. *Jour. AWWA*, 88(3):95-103.

Hart, E.J., K. Sehested, and J. Holcman. 1983. Molar Absorptivities of Ultraviolet and Visible Bands of Ozone in Aqueous Solutions. *Analytical Chemistry*, 55:46-49.

Hatta, S. 1932. Absorption Velocity of Gases by Liquid. II: Theoretical Considerations of Gas Absorption Due to Chemical Reactions. Techn. Repts. Tohoku imp. Univ., 10, 119-135.

Hearn, S.J. 1993. Energy Dissipation Rates and Micromixing in Kenics KMS and Sülzer SMV Motionless Mixer. Internal Report Prepared by BHRA-HILINE.

Heindel, H.L., S.A. Hardy, A. Amirtharajah, and M.J. Arrowood. 1999. Disinfection of *Cryptosporidium parvum* With Static Mixers. In *Proceedings of the American Water Works Association Water Quality and Technology Conference, Oct. 31-Nov. 3, 1999, Orlando, FL.* American Water Works Association, Denver, CO.

Henry, D., E. Freeman. 1994. Finite Element Analysis of Ozone Contactor. In *Proceedings of the International Ozone Association Conference, Pan American Group*, Richmond, VA, 1994.

Herbold, K., B. Flehmig, and K. Botzenhart. 1989. Comparison of Ozone Inactivation, in Flowing Water, of Hepatitis A Virus, Poliovirus 1, and Indicator Organisms. *Applied and Environmental Microbiology*, 55(11):2949-2953.

Hèrés, X. 1993. Etude et Dimensionnement des Mélangeurs Statiques. DEA Génie des procédés, Université de Paris Nord.

Hiatt, C.W. 1964. Kinetics of the Inactivation of Viruses. *Bacteriological Reviews*, 28(2):150-163.

Higbie, R. 1935. The Rate of Absorption of Pure Gas Into a Still Liquid During Short Periods of Exposure. *Trans. Am. Inst. Chem. Eng.*, 35, 360-365.

Hijnen, W.A.M., F.A.P. Houtepen, W.M.H. van der Speld, and D. van der Kooij. 1997. Spores of Sulphite-Reducing Clostridia: a Surrogate Parameter for Assessing the Effects of Water Treatment Processes on Protozoan (Oo)Cysts? In *Proceedings of the International Symposium on Waterborne Cryptosporidium, Newport Beach, CA. March 2-5, 1997.* Edited by C.R. Fricker, J.L. Clancy and P.A. Rochelle. pp. 115-125. American Water Works Association, Denver, CO.

Hishinesvskij, M.K. 1951. The Kinetics of Absorption Under Intense Mixing. *J. Appl. Chem.* USSR, 24, 593.

Hoigné, J., and H. Bader. 1983a. Rate Constants of Reactions of Ozone With Organic and Inorganic Compounds in Water. I: Non-dissociating Organic Compounds. *Water Research*, 17, 173-183.

Hoigné, J., and H. Bader. 1983b. Rate Constants of Reactions of Ozone With Organic and Inorganic Compounds in Water II: Dissociating Organic Compounds. *Water Research*, 17, 185-194.

Huang, C.J., and C.H. Kuo. 1963. General Mathematical Model for Mass Transfer Accompanied by Chemical Reaction. *AIChE J.*, 9, 161-167.

Ishizaki, K., N. Shinriki, and H. Matsuyama. 1986. Inactivation of *Bacillus* Spores by Gaseous Ozone. *Journal of Applied Bacteriology*, 60(67-72).

Jamin-Changeart, T. 1992. Essai sur les Mélangeurs Statiques. Internship report ENSC RENNES, Anjou Recherche.

Kemblowski, Z., and P. Pustelnik. 1988. Residence Time Distribution of Power-Low Fluid in Kenics Static Mixers. *Chem. Eng. Sci.*, 43 (3), 473-478.

Kilani, R.T., and L. Sekla. 1987. Purification of *Cryptosporidium* Oocysts and Sporozoites by Cesium Chloride and Percoll Gradients. *American Journal of Tropical Medicine and Hygiene*, 36:505-508.

Labatiuk, C.W. 1992. Ozone and Ozone-Hydrogen Peroxide Inactivation of *Giardia*. PhD, University of Alberta.

Labatiuk, C.W., F.W. Schaefer, III, G.R. Finch, and M. Belosevic. 1991. Comparison of Animal Infectivity, Excystation and Fluorogenic Dye as Measures of *Giardia muris* Cyst Inactivation by Ozone. *Applied and Environmental Microbiology*, 57(11):3187-3192.

Langlais, C. 1998 Etude des mélangeurs statiques en tant que réacteurs chimique utilisant les procédés d'oxydation O_3/H_2O_2 : approche experimentale et modélisation. PhD thesis University of Rennes 1.

Langlais, B., R. Bonnard, F. Luck, and C. Porte. 1998. Important Parameters for the Ozone Transfer Through Static Mixers. In *Proceedings I.O.A. International Regional Conference on Ozone & AOPs*, 23–25 September, Poitiers (France), 1998.

Langlais, B., D.A. Reckhow, and D.R. Brink, editors. 1991. *Ozone in Water Treatment: Application and Engineering.* Lewis Publishers, Inc., Chelsea, MI.

Lawler, D.F., and P.C. Singer. 1993. Analyzing Disinfection Kinetics and Reactor Design: A Conceptual Approach Versus the SWTR. *Journal of the American Water Works Association*, 85(11):67-76.

Le Sauze, N. 1990. Etude du Transfert de l'Ozone dans les Colonnes à Bulles. Modélisation, Application au Traitement des Eaux Potables et à la Désodorisation de Gaz Soufres. PhD Thesis, Université of Rennes 1, n° d'ordre 458.

Lev, O., and S. Regli. 1992a. Evaluation of Ozone Disinfection Systems: Characteristic Concentration C. *Journal of Environmental Engineering*, 118(4):477-494.

Lev, O., and S. Regli. 1992b. Evaluation of Ozone Disinfection Systems: Characteristic Time T. *Journal of Environmental Engineering*, 118(2):268-285.

Levenspiel, O. 1972. *Chemical Reaction Engineering* Second edition. John Wiley & Sons, Inc., New York, NY.

Levilly, R. 1999. Etude du Mécanisme de Formation des Bromates: Influence de la Température et de la Concentration en Matière Organique. DEA report, University of Rennes and University of Poitiers, France.

Lewis, W.K., and W.G. Whitman (1924) Principles of Gas Absorption. *Ind. Eng. Chem.,* 16, 1215-1220.

Li, H., L.L. Gyürék, G.R. Finch, D.W. Smith, and M. Belosovic 2001. Effect of Temperature on Ozone Inactivation of *Cryptosporidium parvum* in Oxidant Demand-Free Phosphate Buffer. *Journal of Environmental Engineering*, in press.

Lockhart, R.W., and R.C. Martinelli. 1949. Proposed Correlation of Data for Two Phase, Two Component Flow in Pipes. *Chem. Eng. Prog.*, 45 (1):39-48.

Malcolm Pirnie, Inc., and HDR Engineering, Inc. 1991. *Guidance Manual for Compliance with the Filtration and Disinfection Requirements for Public Water Systems Using Surface Water Sources.* Denver, CO: AWWA.

Marinas, J., S. Liang, and E. Aieta. 1993. Modeling Hydrodynamics and Ozone Residual Distribution in a Pilot-Scale Ozone Bubble-Diffuser Contactor. *Journal of the American Water Works Association*, 85 (3), 91-100.

Martin, G., and P. Laffort. 1991. *Odeurs et Désodorisation dans l'Environnement.* Ed Technique et Documentation Lavoisier, Paris, France, 269 p.

Martin, N. 1992. Etude de l'Application de l'Ozone en Traitement des Eaux, Stratégie de Qualification de Cuves Industrielles; Emploi de Mélangeurs Statiques. PhD Thesis, University of Rennes, n° d'ordre 929.

Martin, N., M. Benezet-Toulze, C. Laplace, M. Faivre, and B. Langlais. 1992. Design and Efficiency of Ozone Contactors for Disinfection. *Ozone: Science and Engineering*, 14(5):391-405.

Martin, N., V. Boisdon, A. Laplanche, and P. Uhlig. 1995. Modelization of Industrial Ozonation Tanks: A Useful Tool for the Design of 2000s Ozone Contactor. *Water Supply,* 13(2):57-74.

Martin, N., and C. Galey. 1994. Use of Static Mixer for Oxidation and Disinfection by Ozone. *Ozone Science And Engineering*, 16(6):455-473.

Masschelein, W., G. Fransolet, and J. Genot. 1975. Techniques for Dispersing and Dissolving Ozone in Water /Part 1. *Water and Sewage Works*, 122(12):57-60.

Masschelein, W., G. Fransolet, and J. Genot. 1976. Techniques for Dispersing and Dissolving Ozone in Water / Part 2. *Water and Sewage Works*, 13(2):34-60.

Masschelein, W.J. 1982a. Contacting of Ozone with Water and Contactor Offgas Treatment. In *Handbook of Ozone Technology and Applications, Volume 1.* Edited by R.G. Rice and A. Netzer. pp. 143-224. Ann Arbor, MI: Ann Arbor Science Publishers.

Masschelein, W.J. 1982b. The Direct Action of Dispersed Ozonized Gas Bubbles. In *Ozonation Manual for Water and Wastewater Treatment.* Edited by W. J. Masschelein. pp. 93-95. John Wiley & Sons, New York, NY.

Masschelein, W.J., L. Blaich, E. Thieben, J. Bell, and A. Reading. 1998. Ozone Science and Engineering Special Issue on Quality Assurance in Ozone Practice (1998). *Ozone: Science and Engineering*, 20(6):443-487.

Middleton, J.C. (1978) Motionless Mixers as Gas-Liquid Contacting Devices. AIChE, 71st annual meeting, Miami Beach, USA.

Middleton, J.C. (1985) *Mixing in the Process Industries*. Butterworth & Co.

Munoz Ramirez, G. 1997. Approche cinétique de la demande immédiate en ozone. PhD thesis, University of Rennes 1, order number : 1944.

Myers, K.J., A. Bakker, and D. Ryan. 1997. Avoid Agitation by Selecting Static Mixers. *Chemical Engineering Progress*, 93(6):28-38.

Nauman, E.B. 1982. Reactions and Residence Time Distributions in Motionless Mixers. *Can. J. Chem. Eng*., 60, 136-140.

Nauman, E.B., and B.A. Buffham. 1983. *Mixing in Continuous Flow Systems*. John Wiley and Sons, New York, NY.

Nauman, E.B., and K. Vasudeva. 1985. Residence Time Distribution of Power Low Fluids in Static Mixer. *Can. J. Chem. Eng*., 63, 519-521.

Neter, J., W. Wasserman, and M.H. Kutner. 1989. *Applied Linear Regression Models*. second edition. Irwin, Boston, MA.

Neumann, N.F., L.L Gyürék, G.R. Finch, and M. Belosevic. 2000a. Intact *Cryptosporidium parvum* Oocysts Isolated After in vitro Excystation are Infectious to Neonatal Mice. *FEMS Microbiology Letters,* 186 (1):331-336.

Neumann, N.F., L.L. Gyürék, L. Gammie, G.R. Finch, and M. Belosevic. 2000b. Comparison of Animal Infectivity and Nucleic Acid Staining for Assessment of *Cryptosporidium parvum* Viability in Water. *Appl. Envir. Microbiol.,* 66 (1):406-412.

Nieminski, E.C., and W.D. Bellamy. 1998. Application of Pathogen Surrogates to Improve Treatment Plant Performance. In *Proceedings of the American Water Works Accociation Water Quality Technology Conference, November 5-9, 1998, San Diego, CA*. Denver, CO: American Water Works Association.

Nigam, K.P., and K. Vasudeva. 1980. Residence time distributions in static mixer. *Can. J. Chem. Eng*., 58, 543-544.

Oppenheimer, J., R. Trussel, M. Aieta, J. Jacangelo, and I. Najm. 1999. CT Requirements for Ozone Inactivation for *Cryptosporidium*. AWWA International Symposium on Waterborne Pathogens, August 1999.

Orta de Velasquez, M.T. 1992. Elimination des micropolluants dans les filières d'ozonation du traitement de l'eau potable. Thèse , Université de RENNES I, INSA, N° d'ordre 827.

Owens, J.H., R.J. Miltner, E.W. Rice, C.H. Johnson, D.R. Dahling, F.W. Schaefer, and H.M. Shukairy. 2000. Pilot-scale ozone inactivation of *Cryptosporidium* and other microorganisms in natural water. *Ozone: Science and Engineering*, 22(5):501-517.

Perrot, C. 2000. Influence de Paramètres Physico-Chimiques de l'Eau sur la Formation des Radicaux Hydroxyls : Tentative de Modélisation, Rapport DEA de Chimie et Microbiologie de l'Eau.

Rader, R.G., M. Mutsakis, F. Grosz-Röll, and W. Maugweiler. 1989. Better Absorption? Try a Static Mixer. *Chemical Engineering*, 96(7):137-142.

Ramirez, C.R. (1997) Contribution à l'Etude du Comportement Hydrodynamique des Contacteurs d'Ozonation: Influence de la Géométrie et des Paramètres Opératoires. PhD Thesis. INSA Toulouse.

Reddy, Y.R., and S. Kar. 1968. Theory and Performance of Water Jet Pumps. *In Proceedings of the American Society of Civils Engineers*, HY, 1261-1283.

Reddy, S., R. Joost, and J. Heckler. 1995. Ozone Contactor Hydrodynamics. *Proceedings of the 12th Ozone World Congress*, Lille, France, pp. 175-187.

Rice, E.W., K.R. Fox, R.J. Miltner, D.A. Lytle, and C.M. Johnson. 1996. Evaluating Plant Performance with Endospores. *Journal of the American Water Works Association*, 88(9):122-130.

Richards, D.A., and M. Fleischman. 1975. Ozone Transfer to Aqueous Systems in a Static Mixer. In *Proceedings of the Second International Symposium on Ozone Technology.* Edited by R.G. Rice, P. Pichet and M.-A. Vicent. pp. 57-70. Montreal, Canada:

Roberts-Thomson, I.C., D.P. Stevens, A.A.F. Mahmoud, and K.S. Warren. 1976. Giardiasis in the Mouse: An Animal Model. *Gastroenterology*, 71:57-61.

Roes, A.W.M., A.T. Zeeman, and F.H.J. Bukkems. 1984. High-Intensity Gas-Liquid Mass Transfer in the Bubbly Flow Region During Co-current Upflow Through Static Mixers. I. Chem. E.; Symp. Ser. N° 38, 231-238.

Roustan, M., H. Debellefontaine, Z. Do-Quang, and J.P. Duguet. 1997. Development of a Method for the Determination of Ozone Demand of a Water. In *Proceedings of the IOA 13th Ozone World Congress*, Kyoto, October 26-31.

Roustan, M., J.H. Kim, and B.J. Marinas. 2000. Hydrodynamics of Ozone Reactors. IOA symposium "fundamental and engineering concepts for ozone reactor design". Toulouse, France, March 1-3.

Roustan, M., C. Beck, O. Wable, J.P. Duguet, and J. Mallevialle. 1993. Modeling Hydraulics of Ozone Contactors. *Ozone: Science and Engineering*, 15(3):213-226.

Roustan, M., R.Y. Wang, and D. Wolbert. 1994. Modeling of Ozone Mass Transfer in a Bubble Column: Determination of the Mass Transfer Coefficient Kla and the Consumption of Ozone by Water. In *Proceedings IOA,* Warsaw, Poland.

Roustan, M., R.Y. Wang, and D. Wolbert. 1996. Modeling Hydrodynamics and Mass Transfer Parameters in a Continuous Ozone Bubble Column. *Ozone: Science and Engineering*, 18 (2), 99-115.

Russel, W., R.P. Hesketh, and A.W. Etchells. 1987. Bubble Size in Horizontal Pipelines. *AIChE*, 33.

Savary, B., R. Levilly, A. Laplanche, D Wolbert., C. Galey, and J. Cavard. 2000. Formation of Hydroxyl Radicals: Influence of the Water Characteristics. In *Proceedings IOA, Wasser Berlin 2000 Conference*, October 24 -27.

Scaccia, C., and H.M. Rosen. 1978. Ozone Contacting: What is the Answer? *Ozonews*, 5(10, Pt. 2):1-8.

Scharfenaker, M.A. 2000. Water Suppliers Assess New Rulemaking Agreement. *Journal of the American Water Works Association*, 92 (11), 22-33

Schulgen, B., A. Amirtharajah, and S.C. Jones. 1996. Effectiveness of Static Mixers for Coagulation in Water Treatment. In *Proceedings of the American Water Works Association Annual Conference, June 23-27, Toronto, Ont.* pp. 121-147. American Water Works Association, Denver, CO.

Schulz, C.R., and P.W. Prendiville. 1993. Designing High Concentration Ozone Contactors for Drinking Water Treatment Plants. *Ozone: Science and Engineering*, 15, 245-266.

Schulz, C.R., G.C. Shafran, L.B. Garrett, and R. Hawkins. 1995. Evaluating a High-Efficiency Ozone Injection Contactor. *Journal of the American Water Works Association*, 87(8):85-99.

Seber, G.A.F., and C.J. Wild. 1989. *Nonlinear Regression.* John Wiley & Sons, New York, NY.

Shah, N.F., and D.D. Kale. 1992. Pressure Drop for Laminar Flow of Non Newtonian Fluids in Static Mixers. *Chem. Eng. Sci.*, 46 (8), 2159.

Smith, D.W., and H. Zhou. 1993. Theoretical Analysis of Ozone Disinfection Performance in a Bubble Column. *Proceedings of the 11th Ozone World Congress*, San Francisco.

Song, R., P. Westerhoff, R. Minear, and G. Amy. Bromate Minimization During Ozonation. 1997. *Jour. AWWA,* 89:6:69-78.

Staehelin, J., and J. Hoigné. 1982. Decomposition of Ozone in Water: Rate of Initiation by Hydroxides Ions and Hydrogen Peroxide. *Environ. Sci. Technol.*, 16(10):676-681.

Stahelin, J., R.E Bülher, and J. Hoigné. 1984. Ozone Decomposition in Water Studied by Pulse Radiolysis. II: OH° and HO4° as Chain Intermediates. *J. Phys. Chem.*, 22 (24): 5999-6004.

Steiff, F.A., and F. Käser. 1991. Sülzer Mixer Reactor SMR for Gas/Liquid Reactions. 7th European Conference on Mixing, Bruges, Belgium.

Studer, M. 1991. Residence Time Distribution and Pressure Drop Measurements in the Kenics Mixer. Internal Report Prepared by BHRA-HILINE.

Toenniessen, G.H., and D.J. Johnson. 1970. Heat Shocked *Bacillus subtilis* Spores as an Indicator of Virus Disinfection. *Journal of the American Water Works Association*, 62(9):589-593.

Tomiak, R.B., J.L. Rennecker, and Mariñas. 1998. Modeling of *Cryptosporidium parvum* Oocyst Inactivation in a Pilot-Scale Bubble-Diffuser Ozone Contacter. In AWWA *Water Quality and Technology Conference*. San Diego, CA: AWWA.

Toor, H.L., and J.M. Marchello. 1958. Film-Penetration Model for Mass and Heat Transfer. *AIChE J.*, 4, 97-101.

Trussell, R.R., and J.-L. Chao. 1977. Rational Design of Chlorine Contact Facilities. *Journal of the Water Pollution Control Federation*, 49(4):659-667.

Tung, T.T. (1976) Low Reynold Number Entrance Flow: a Study of a Motionless Mixer. Ph D Thesis, University of Massachusetts, USA.

Turunen, I., and H. Haario. 1994. Mass Transfer in Tubular Reactors Equipped with Static Mixers. *Chem. Eng. Sci.,* 49 (24):5257-5269.

Van Krevelen, D.M., and P.J. Hoftijser. 1948. Kinetics of Gas Liquid Reactions Part 1: General Theory. *Receuils Trav. Chem.*, Netherlands, vol. 67, pp. 563.

Vernon, N.A. 1995. Effect of Viscosity on Gas-Liquid Mixing and Mass Transfer in Kenics Mixer. Internal Report Prepared by BHRA-HILINE.

Villermaux, J. 1993. *Génie de la Réaction Chimique*, Lavoisier, Paris, France.

Wang, R.Y. (1995a) Etude de l'hydrodynamique et du transfert de l'ozone dans une colonne à bulles en ascension libre. PhD Thesis, Université de TOULOUSE, INSA, N° d'ordre 341.

Wang, R.Y. (1995b) Axial dispersion models of fluid-fluid reactors based on the two film theory. *Chem. Eng. and Proc.*, 34 (5), 447-453.

Wang, J., and H. Langemann. 1994a. Unsteady Two Flow Model for Mass Transfer. *Chem. Eng. Technol.*, 17, 280-284.

Wang J., and H. Langemann. 1994b. Unsteady Two Flow Model for Mass Transfer Accompanied by Chemical Reaction. *Chem. Eng. Sci.*, 49, 3457-3,463.

Whitman, W.G. (1923) The two-film theory of gas absorption. *Chem. Metall. Eng.*, 29, 146-148.

Wickramanayake, G.B., A.J. Rubin, and O.J. Sproul. 1984. Inactivation of *Naegleria* and *Giardia* Cysts in Water by Ozonation. *Journal of the Water Pollution Control Federation*, 56(8):983-988.

Yurteri, C., M.D. Gurol. 1988. Ozone consumption in natural waters: effects of background organic matter, pH, carbonate species. *Ozone: Science and Engineering.*, 10, 277-290.

Zhou, Z.M. 1991. Development of Two Phase Measurement Techniques and Studies of Mass Transfer of a Kenics Mixers. Internal Report Prepared by BHRA-HILINE.

Zhou, Z.M., J. Hannon, and A. Green. 1992. Use of High Intensity Gas-Liquid Mixers as Reactors. *Chem. Eng. Sci.*, 47 (9-11):2847-2852.

Zhu, Q. 1990. Ozone Disinfection Of Sewage In A Static Mixer Contacting Reactor System On A Plant Scale. *Ozone: Science and Engineering*, 13(3):313-328.

Zhu, Q., C. Liu, and Z. Xu. 1989a. Study of Contacting Systems in Water and Wastewater Disinfection by Ozone. 1. Mechanism of Ozone Transfer and Inactivation Related to the Contacting Method Selection. *Ozone: Science and Engineering*, 11(2):169-188.

Zhu, Q., C. Liu, and Z. Xu. 1989b. Study of Contacting Systems in Water and Wastewater Disinfection by Ozone. 2. Mathematical Models of the Ozone Disinfection Process With a Static Mixer. *Ozone: Science and Engineering*, 11(2):189-207.

Zhu, Q. 1991. Ozone Disinfection of Sewage In a Static Mixer Contacting Reactor System on a Plant Scale. *Ozone: Science and Engineering OZSEDS,* 13(3):313-330.

ABBREVIATIONS AND SYMBOLS USED

a	superficial area per volumetric unit
A_{260}	absorbance measured at 260 nm in a 10 mm cell
A_{600}	absorbance measured at 260 nm in a 10 mm cell
ATCC	American Type Culture Collection
AWWA	American Water Works Association
AwwaRF	Awwa Research Foundation
°C	degrees Celsius
C_a	applied ozone represented as concentration, mg/L
C_{avg}	average ozone concentration, mg/L
$C_{avg}t_b$	average ozone concentration × time product in the batch reactor
$C_{avg}t_m$, Ct	average ozone concentration × time product in the static mixer contactor
C_f	final ozone concentration, mg/L
C_0	initial ozone concentration in batch reactor or at inlet of reactive flow segment in static mixer contactor, mg/L
CFSTR	continuous-flow stirred tank reactor
CFU	colony forming fnit
CWD	Cleveland Water Department
d	number of live oocysts in inoculum estimated from infectivity model
d_0	total number of oocysts in inoculum based on hemocytometer counts
d_{32}	Sauter diameter
$D_{A,g}$	diffusion coefficient of the component A in gas phase
$D_{A,l}$	diffusion coefficient of the component A in liquid phase
d_b	bubble diameter
d_h	hydraulic diameter
DBP	disinfection by-product
DOC	dissolved organic carbon
$E(t)$	exit age distribution determined from tracer tests
E_a	activation energy
EWC	Elizabethtown Water Company
f	fraction of the spore population with a higher resistance to ozone
FBD	fine bubble diffuser
FBDU	fine bubble diffuser pilot unit
GAC	granular activated carbon
GC	Gwinnett County
G/L	gas to liquid flow rate ratio
gpm	gallon per minute

H	height of the gas/liquid separator in the Laboratory Study apparatus
Ha	Hatta criterion
He'	Henry constant
HPC	heteretrophic plate count
HRT	hydraulic residence time
ID	internal diameter
ID_{50}	parasite dose required for a 50% rate of infection
IESWTR	Interim Enhanced Surface Water Treatment Rule
in.	inch
k	rate constant of the Hom kinetic model, natural logarithm
k'	rate constant of the Hom kinetic model, base 10 logarithm
k_b	rate constant for inactivation of spores with normal ozone resistance
k_d	first order decay coefficient for ozone decomposition
k_La	mass transfer coefficient
k_r	rate constant for inactivation of spores with higher ozone resistance
L	water flow rate, L/min
LADWP	Los Angeles Department Water and Power
Lh	likelihood function
L/min	liter per minute
LP	latent period of infection
l_S	length of a static mixing element
l_{SM}	global length of the static mixer
LT2ESWTR	Long Term 2 Enhanced Surface Water Treatment Rule
l_V	length of the void between two elements
m	kinetic parameter of the Hom kinetic model of ozone inactivation
MCL	maximum contaminant level
mg/L	milligram per liter
mL	milliliter
m/s	meter per second
μg/L	micrograms per liter
N	concentration of live microorganisms after a specified ozone treatment
n	kinetic parameter of the Hom kinetic model of ozone inactivation
N_0	concentration of live microorganisms before ozone treatment
N_A	transfer flux of the component A
N_b	concentration of spores with normal resistance to ozone
n_c	number of critical targets required for spore inactivation
N_r	concentration of spore with higher resistant to ozone
N.A.	not available
NOM	natural organic matter
NTU	nephelometric turbidity units

OD	ozone dose
P	proportion of neonatal CD-1 mice that are positive for infection
P_A^*	partial pressure of the component A at the interface gas-liquid
$P_{A,e}$	partial pressure of the component A at equilibrium
PCTE	polycarbonate track-etched
PFA	plug flow analysis
PFR	plug flow reactor
pKa	dissociation kinetic constant
P_{out}	pressure at the outlet of the static mixer
p-value	probability of a type I error in a hypothesis test
PVC	polyvinyl chloride
R	universal gas constant
Re	Reynold's number
RFS	reactive flow segment
RTD	residence time distribution
s	second
SE	standard error
SFA	segregated flow analysis
SM	static mixer
SMU	static mixer unit
SWTR	surface water treatment rule
t_{10}	time required for 10% of the injected substance to exit the system
t_{90}	time required for 90% of the injected substance to exit the system
t_b	batch contact time
t_m	mean residence time
TE	transfer efficiency
TOC	total organic carbon
UV254	ultraviolet light absorbance at 254 nm wavelength
V_d	dispersion volume of the static mixer device
v_s	liquid velocity through the static mixer
We	Weber number
WTP	water treatment plant
Y_f	concentration of ozone in contactor feed gas (mg/L or g/m^3)
Y_o	concentration of ozone in contactor off-gas (mg/L or g/m^3)

SYMBOLS

β_0	parameter of the *Cryptosporidium parvum* logit dose-response model
β_1	parameter of the *Cryptosporidium parvum* logit dose-response model
δ_g	layer thickness of the gas interface
δ_l	layer thickness of the liquid interface
ΔP_g	gas pressure drop
ΔP_l	liquid pressure drop
ε_g	gas holdup
ε_l	liquid holdup
γ	the incomplete gamma function
σ	interfacial tension

IWA Publishing's authorised EU representative for General Product Safety Regulations is Diane D'Arras, 15 rue Duret, 75116 Paris, France, e-mail: safety@iwap.co.uk.

Printed and bound by CPI Group (UK) Ltd, Croydon, CR0 4YY
17/09/2026
02218813-0004